165

Advances in Polymer Science

Springer

Berlin
Heidelberg
New York
Hong Kong
London
Milan
Paris
Tokyo

Polyelectrolytes with Defined Molecular Architecture I

Volume Editor: Manfred Schmidt

With contributions by
M. Ballauff · M. Biesalski · J. Bohrisch · P. Dziezok
C.D. Eisenbach · H. Engelhardt · M. Eschner
K. Estel · F. Gröhn · C.A. Helm · N. Houbenov
N. Hugenberg · W. Jaeger · D. Johannsmann
R.v. Klitzing · R. Konradi · M. Martin · H. Menzel
T. Meyer · S. Minko · M. Motornov · H. Mori
H. Möhwald · A.H.E. Müller · R.R. Netz · D. Pleul
M. Rehahn · J. Rühe · C. Schaller · M. Schmidt
C. Seidel · F. Simon · S. Spange · M. Stamm
T. Stephan · B. Tieke · S. Traser · D. Usov · I. Voigt
P. Wittmeyer · H. Zhang

 Springer

The series presents critical reviews of the present and future trends in polymer and biopolymer science including chemistry, physical chemistry, physics and material science. It is addressed to all scientists at universities and in industry who wish to keep abreast of advances in the topics covered.

As a rule, contributions are specially commissioned. The editors and publishers will, however, always be pleased to receive suggestions and supplementary information. Papers are accepted for "Advances in Polymer Science" in English.

In references Advances in Polymer Science is abbreviated Adv Polym Sci and is cited as a journal.

The electronic content of APS may be found at http://www.SpringerLink.com

ISSN 0065-3195
ISBN 3-540-00528-5
DOI 10.1007/b10953
Springer-Verlag Berlin Heidelberg New York

Library of Congress Catalog Card Number 61642

Springer-Verlag Berlin Heidelberg New York
Springer Verlag is a part of Springer Science+Business Media
springer.de

Typesetting: Stürtz AG, 97080 Würzburg
Cover: Design & Production, Heidelberg
Printed on acid-free paper   02/3020/kk – 5 4 3 2 1 0

Volume Editor

Prof. Dr. Manfred Schmidt
Institut für Physikalische Chemie
Universität Mainz
55128 Mainz, Germany

Editorial Board

Advances in Polymer Science
Available Electronically

For all customers with a standing order for Advances in Polymer Science we offer the electronic form via SpringerLink free of charge. Please contact your librarian who can receive a password for free access to the full articles. By registration at:

http://www.SpringerLink.com

If you do not have a standing order you can nevertheless browse through the table of contents of the volumes and the abstracts of each article by choosing Advances in Polymer Science within the Chemistry Online Library.

You will find information about the

– Editorial Board
– Aims and Scope
– Instructions for Authors
– Sample Contribution

at www.springeronline.com using the search function.

Preface

Back in 1996 the German Science Foundation (Deutsche Forschungsgemein-schaft) has launched a nationwide research center on "Polyelectrolytes with defined molecular architecture–synthesis, function and theoretical description" (DFG-Schwerpunkt-Programm 1009: Polyelektrolyte mit definierter Molekül-architektur–Synthese, Funktion und theoretische Beschreibung). On average 25 research groups from all over Germany and one French group were funded for a total of six years in order to attack and solve long standing problems in the field, to explore new ideas and to create new challenges.

The scientific achievements of this center of research are summarized in the present volumes of Advances in Polymer Science, volume 165 and 166. Financially supported by a "Coordination Funds" the interdisciplinary cooperation between the very many participating research groups was greatly enhanced and has consequently led to contributions involving an unusually large number of authors.

We hope that the center has brought German Polyelectrolyte Research into an international leading position and that it will constitute the nucleus for future activities in this field.

On behalf of all of my colleagues I wish to thank the "Deutsche Forschungs-gemeinschaft" for financial and in particular Dr. K.-H. Schmidt and Dr. F.-D. Kuchta for administrative support and to the voluntary reviewers of the proposals, Prof. Blumen, Univ. Freiburg, Prof. Fuhrmann, Univ. Clausthal, Prof. Heitz, Univ. Marburg, Prof. Maret, Univ. Konstanz, Prof. Möller, Univ. Ulm, Dr. Winkler, BASF Ludwigshafen, Prof. Wulf, Univ. Düsseldorf, for their invaluable judgment and advice.

Mainz, February 2003   Manfred Schmidt

Contents

Adv Polym Sci (2004) 165:1–41
DOI 10.1007/b11266

New Polyelectrolyte Architectures

Jörg Bohrisch[1] · Claus D. Eisenbach[2] · Werner Jaeger[1] · Hideharu Mori[3] ·
Axel H. E. Müller[3] · Matthias Rehahn[4] · Christian Schaller[2] · Steffen Traser[4] ·
Patrick Wittmeyer[4]

[1] Fraunhofer Institut für Angewandte Polymerforschung, 14476 Golm, Germany
  E-mail: *werner.jaeger@iap.fraunhofer.de*
[2] Forschungsinstitut für Pigmente und Lacke e.V., 70569 Stuttgart und Institut für
  Angewandte Makromolekulare Chemie, Universität Stuttgart, 70569 Stuttgart, Germany
  E-mail: *cde@fpl.uni-stuttgart.de*
[3] Makromolekulare Chemie II und Bayreuther Zentrum für Kolloide und Grenzflächen,
  Universität Bayreuth, 95440 Bayreuth, Germany
  E-mail: *Axel.Mueller@uni-bayreuth.de*
[4] Institut für Makromolekulare Chemie, Technische Universität Darmstadt,
  64287 Darmstadt, Germany
  E-mail: *MRehahn@dki.tu-darmstadt.de*

Abstract This chapter reviews recent advances in the synthesis of polyelectrolytes. New results are presented for linear polymers with stiff and flexible backbones and with star-shaped and randomly branched structures. Block and graft copolymers containing both charged ionic and hydrophobic monomer units are also discussed. Homo- and block copolymers of carbobetaines complete the variation of molecular architecture. Synthetic approaches mainly use anionic, controlled radical, and macromonomer polymerization techniques, and the Suzuki reaction, to synthesize reactive precursors which are subsequently transformed into the final products. Results from physicochemical characterization of the new polymers are also mentioned.

Keywords Polyelectrolytes · Polycarbobetaines · Block copolymers · Graft copolymers · Reactive precursors

1
Introduction

Polyelectrolytes play a major part in nature and find manifold application in industrial processes and daily life. Because they are both macromolecules and electrolytes they represent a unique class of polymer. The special properties of these polymers are determined by their electrochemical and macromolecular parameters and their chemical structure. Charge density and charge strength result in variable long-range electrostatic interactions, which are mainly responsible for the physicochemical properties of polyelectrolytes and their peculiarities in aqueous solution. These interactions and the formation of complexes driven by attractive Coulomb forces between charged macromolecules and oppositely charged macroions, surfactants, colloid particles, or solid surfaces are of central importance in the life sciences and in most technical applications. However, the typical properties of polyelectrolytes do not depend exclusively on the electrostatic forces. Differences between the flexibility of the polymer chain and, especially, a molecular architecture supporting the formation of H-bonds or hydrophobic interactions also play a major role in questions of scientific interest and practical importance.

Due to the comparatively high structural variability of polyelectrolytes an increasing number of applications with high practical relevance has been established, but up to now the fundamentals of many of these processes are not well understood. Also many topics in the life sciences, where polyelectrolytes are involved, bear similar open questions. To support any research in this field tailor-made polyelectrolytes with well defined molecular architecture and molecular parameters are required. Additionally, new polymers with adjusted properties may contribute to the optimization of known applications of polyelectrolytes and to the opening of new applications.

This chapter deals with the synthesis and properties of polyelectrolytes bearing a new molecular architecture. The following structures are included:

- rodlike poly(*p*-phenylene) polyelectrolytes
- ionically charged block copolymers and polycarbobetaines
- nonlinear polyelectrolyte topology
- amphipolar graft copolyelectrolytes.

2
Synthesis of Rodlike Poly(*p*-phenylene) Polyelectrolytes

To tailor-make polyelectrolytes for a specific aim, profound understanding of their structure–property relationships is required. This understanding, however, is not yet available, despite intense research during recent decades [1–3]. To fill the gaps in our knowledge, comprehensive consideration of:

1. the intramolecular electrostatic forces,
2. their intermolecular counterparts,
3. the osmotic effects, and
4. possible conformational changes occurring in aqueous polyelectrolyte solutions

is necessary. All these parameters have to be correlated with the chain's architecture and the composition of the respective solution. Unfortunately, such an analysis is very difficult for flexible polyelectrolytes. In these systems, both the intramolecular and intermolecular coulomb forces depend on the ionic strength. Hence the coil dimensions and the range of intermolecular electrostatic repulsion change simultaneously with changing ionic strength, but according to unknown rules. As a consequence, the different parameters contributing to the macroscopically observed polyelectrolyte behavior cannot be determined individually. Thus it is hardly possible to develop the required theoretical understanding just by analyzing flexible systems. Conformationally rigid, rodlike polyelectrolytes, on the other hand, are much easier to understand because they do not change their shape as a function of the ionic strength. Since conformational changes can be neglected here, all effects observed on changing the ionic strength can be attributed to electrostatics. Once this special aspect of polyelectrolyte behavior of rodlike systems is understood, it should be much easier to describe flexible polyelectrolytes also when conformational changes have to be taken into account. Therefore rodlike polyelectrolytes are considered to be ideal model systems for developing the required full understanding of polyelectrolytes in solution.

Rodlike polyelectrolytes have been known for a long time. Biological polymers such as DNA [4–8] and xanthane [9–12], or colloidal systems like the ferredoxin virus [13, 14] and the tobacco mosaic virus (TMV) [15, 16], may be the most prominent examples. However, there were also publications on some synthetic rods in the early 1990s when we started our program. Poly(*p*-phenylene-benzobisoxazoles) and poly(*p*-phenylene-benzobisthiazoles) may serve as examples [17–20]. Nevertheless, new rodlike polyelectrolytes were developed based on poly(p-phenylene) (PPP). This step seemed to be necessary because a system was required which is chemically very stable and rodlike under all the conditions its aqueous solution may be exposed to. PPP fulfils these requirements as it is:

1. intrinsically rodlike (persistence length $l_p \approx 20$ nm [21]) due to its all-*para*-linked phenylene repeating units, and
2. perfectly inert against hydrolysis and all other reactions possible in aqueous media.

Moreover, PPP was selected because the powerful Pd-catalyzed Suzuki coupling reaction in combination with the concept of solubilizing side chains offers many advantages in synthesis.

2.1
PPP Polyelectrolytes via Ether Intermediates

Due to the pronounced tolerance of the Suzuki reaction towards additional functional groups in the monomers, precursor strategies as well as so called direct routes can be applied for polyelectrolyte synthesis. However, the latter possibility, where the ionic functionalities are already present in the monomers, was rejected. The reason is too difficult determination of molecular information by means of ionic polymers. Therefore the decision was to apply precursor strategies (Scheme 1). Here, the Pd-catalyzed polycondensation process of monomers **A** leads to a non-ionic PPP precursor **B** which can be readily characterized. Then, using sufficiently efficient and selective macromolecular substitution reactions, precursor **B** can be transformed into well-defined PPP polyelectrolytes **D**, if appropriate via an activated intermediate **C**.

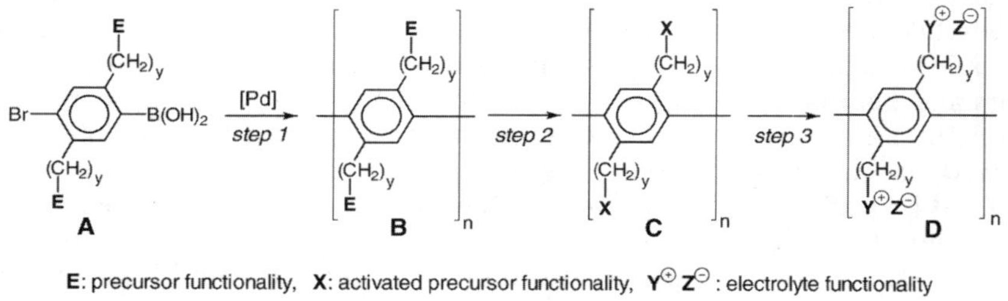

E: precursor functionality, X: activated precursor functionality, $Y^{\oplus} Z^{\ominus}$: electrolyte functionality

Scheme 1 General representation of precursor syntheses leading to PPP polyelectrolytes

For most of the polymers, ether groups **E** were selected as precursor functionalities because they are inert under the conditions of the Pd-catalyzed polycondensation reaction but can be converted selectively and completely into reactive groups $-(CH_2)_y-X$ such as alkyl bromides or alkyl iodides that allow the final introduction of the ionic functionalities.

In a first sequence of experiments, butoxymethylene-substituted precursor PPPs **3** were prepared (Scheme 2) [22, 23]. The polycondensation of equimolar amounts of **1** and **2** leads to constitutionally homogeneous products **3** having values of $P_n \approx 60$. The lateral benzylalkyl ether groups were then cleaved quantitatively, leading to the nicely soluble bromomethylene-functionalized activated precursor **4** which was finally converted into, for example, the carboxylated PPP**5**. Unfortunately, this polymer proved to be insoluble in water or aqueous bases. It was reasonable to assume that this is due to the relatively low density of charged groups along the chains, and also to the apolar alkyl side chains attached to every second phenylene moiety causing intermolecular hydrophobic interactions. An attempt was therefore made to make the corresponding homopolymer **7** available. Under very specific conditions it was possible to obtain the required AB type monomer **6**. After successful polycondensation, however, it was no longer possible to transfer precursor PPP **7** into polyelectrolyte **9**. Due to the lack of solubilizing side chains in the activated intermediate **8**, this material was insoluble and could not be converted into a constitutionally homogeneous product.

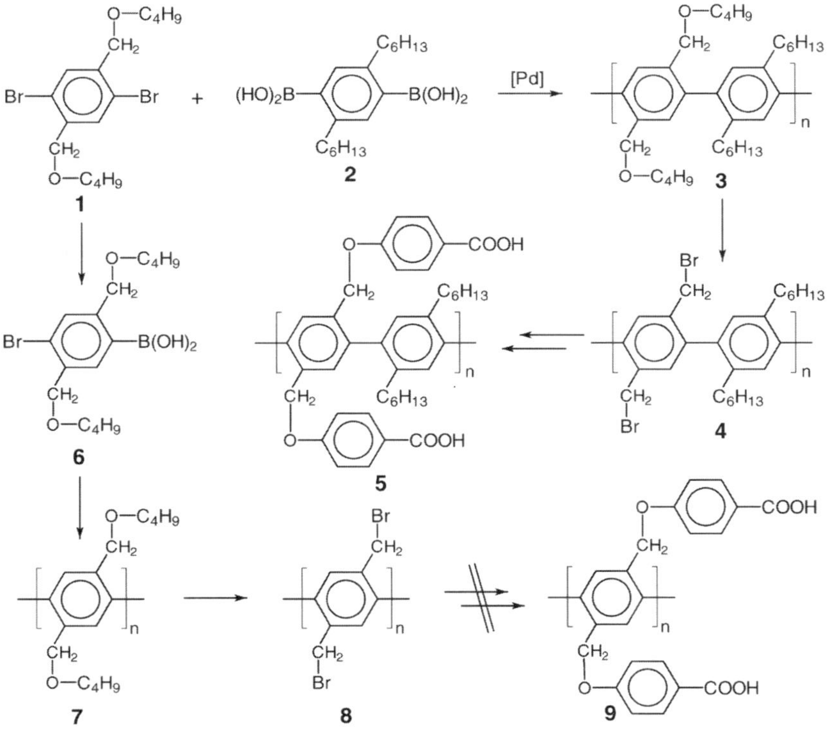

Scheme 2 Precursor synthesis of carboxylated PPP polyelectrolytes

This failure forced a change in the synthetic strategy. The two functions of the two different lateral substituents of **3**, i.e. solubilizing the polymer (done by the C_6H_{13} groups) and making possible final introduction of electrolyte functionalities (done by the CH_2–O–C_4H_9 groups), were combined in one single type of side chain [24]. This was realized by introducing a larger spacer group between the PPP main chain and the ether functionality. Hexamethylene spacers were sufficient to solubilize the rodlike macromolecules even in the activated state. Due to the longer spacers, however, the ether functionality was no longer in a benzylic position. To nevertheless maintain the selectivity of the ether cleavage process, which must guarantee the formation of 100% halogenalkyl groups and 0% hydroxy groups in the activated polymer–to prevent crosslinking–alkyl phenyl ethers were used as the precursor's functionality.

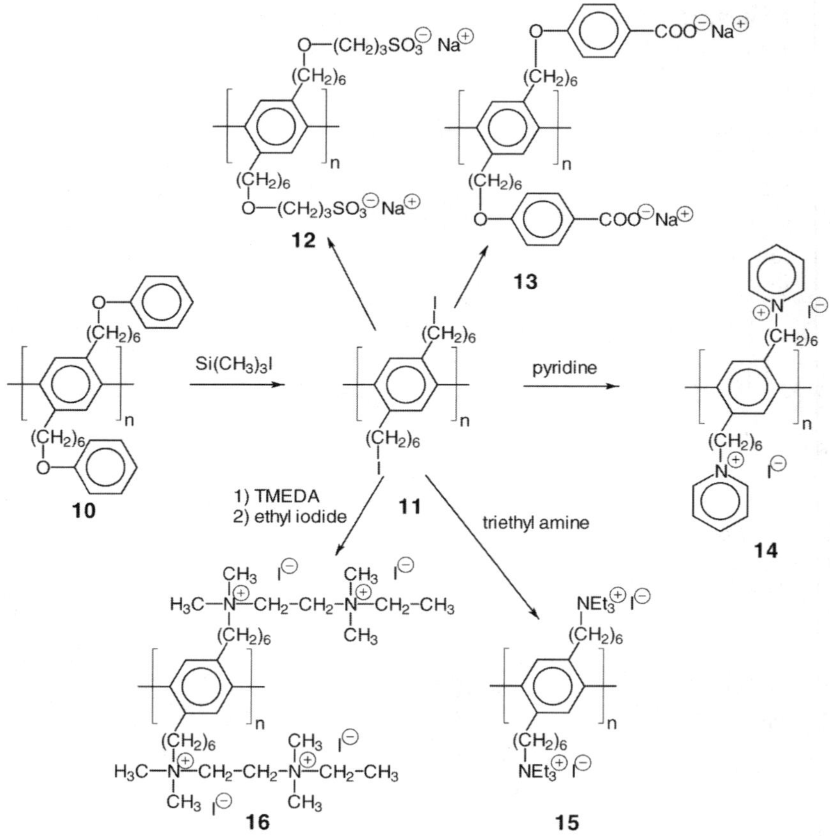

Scheme 3 PPP polyelectrolytes available from precursor **10**

The synthesis of the required monomer and of precursor polymer **10** proved to be time-consuming but not difficult. The subsequent ether cleavage **10**→**11** using $(H_3C)_3$ Si–I in CCl_4 occurs very homogeneously, and also completely, if strictly water-free conditions are kept (Scheme 3). Rather surprisingly, however, all anionic polyelectrolytes prepared from **11**, such as **12** and **13**, proved to be insoluble in water or aqueous bases, despite their charge density being twice as high as in polymers like **5** [24]. In contrast to this, cationic polyelectrolytes such as **14–16**, easily available via conversion of **11** with a tertiary amine, proved to be molecular-dispersal soluble not only in polar organic solvents but even in pure water [25, 26]. It is believed that this is because the apolar interior of these cylinder-like polyelectrolytes is covered by a sufficiently "dense" shell of hydrophilic cationic groups which prevent intermolecular hydrophobic interactions. In the case of anionic polyelectrolytes such as **12** and **13**, on the other hand, the density of the charged groups in the cylinder shell is not high enough due to the longer spacers between the main chain and the electrolyte functionality. Therefore, they are insoluble in water.

Using polyelectrolytes **14–16**, a huge number of studies have been performed by means of, for example, viscosimetry, membrane osmometry, small-angle X-ray scattering (SAXS), and electric birefringence. Because selected results are described in the chapter "Stiff-Chain Polyelectrolytes", we only report here on the key results of viscosity experiments. Salt-free aqueous solutions of the above rodlike polyelectrolytes always display a polyelectrolyte effect which is much more pronounced than that of flexible-chain analogs of comparable chain length. Moreover, the maximum value of η_{sp}/c_P always appears at values of c_P approx., one order of magnitude lower than with flexible systems. When salt is added to these solutions, the maximum becomes weaker and shifts towards higher values of c_P. Finally, at salt concentrations higher than $c_S = 2 \times 10^{-4}$ mol L^{-1}, a linear Huggins plot is obtained which gives an intrinsic viscosity $[\eta]$ nicely corresponding to that of the precursor polymer used for the preparation of the respective polyelectrolyte.

SAXS and osmometry, on the other hand, allow the conclusion that the Poisson–Boltzmann cell model gives a quite realistic description of counterion condensation in rodlike macromolecules. However, prior to a final evaluation, a more profound analysis is required. Here, it will be of particular importance to consider polyelectrolytes with substantially lower charge densities also. Unfortunately, but in accordance with expectations, all polyelectrolytes containing phenylene moieties without charged side groups, such as **20–22**, proved to be insoluble in water (Scheme 4).

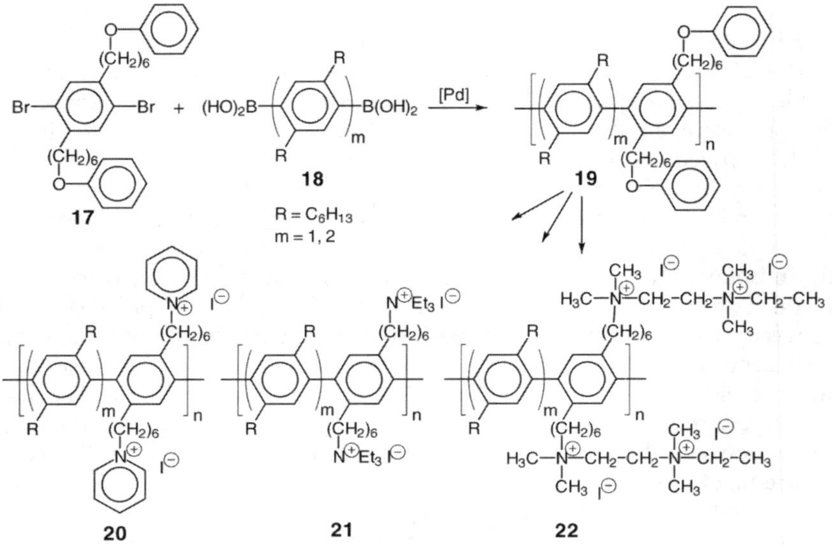

Scheme 4 Cationic PPP polyelectrolytes having different charge densities

2.2
PPP Polyelectrolytes via Amino Intermediates

Because of the findings reported above, the development of a new strategy was recently started which should make available PPPs which are soluble in water even at vanishing charge density. The key step of this new strategy is the attachment of oligoethylene oxide (OEO) side chains to the PPP backbone [27]. These substituents should allow:

1. avoidance of the apolar alkyl side chains and, moreover,
2. efficient prevention of hydrophobic interactions of the apolar PPP main chains.

The change from alkyl to OEO substituents as solubilizing side chains, however, has far-reaching consequences for the precursor strategy–OEO side chains do not allow use of the procedure used so far where ether cleavage is a key step. The most convenient way to avoid ether cleavage as a macromolecular substitution step is to invert the above synthetic strategy, i.e. to use tertiary amines as the precursor functionalities and to generate the polyelectrolyte via treatment of the precursor with low-molecular-weight alkyl halides (Scheme 5).

A positive side-effect of the new strategy is a clearly simplified and accelerated polyelectrolyte synthesis: the monomers are quickly available and the time-consuming ether cleavage is no longer necessary. Despite the fact that not all desired materials could yet be obtained, we nevertheless can recon-

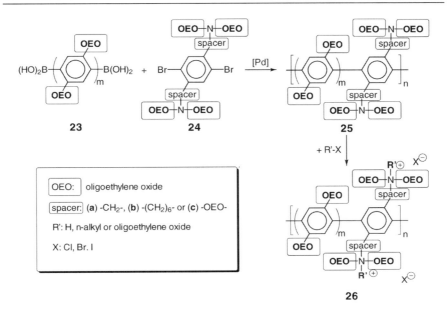

Scheme 5 General representation of the synthesis of cationic PPP polyelectrolytes via water-soluble amino-functionalized precursor PPPs

firm that the new concept indeed works. First of all it was found out–by means of extensive model investigations–under which conditions the Pd-catalyzed polycondensation proceeds unaffected by the polar, potentially coordinating, side chains.

Based on this knowledge the first uncharged but nevertheless readily water-soluble precursor PPP **27** ($P_n \approx 30$) were recently obtained. As the degree of protonation may play an important role in the profound characterization of polyamines like **27**, titration studies were also performed (Scheme 6).

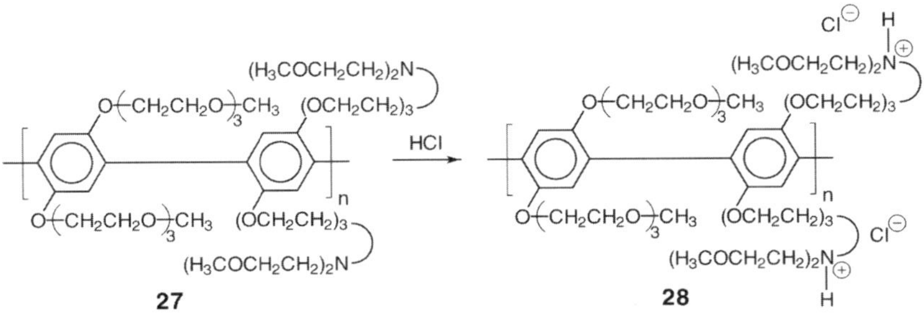

Scheme 6 Protonation of precursor PPP **27**

The experiments show that the degree of protonation can be monitored using either ^{1}H NMR spectroscopy or potentiometry. The lower plot in Fig. 1 shows the degree of protonation of polymer **27** as a function of pH. For comparison, the upper plot in Fig. 1 displays the corresponding picture for its low-molecular-weight analog N[(CH$_2$CH$_2$O)$_2$CH$_3$]$_3$. A buffer region between pH $\approx$ 8 and pH $\approx$ 6 is evident in both cases. However, the precursor PPP seems to be the weaker base. The solid lines between 10% and 90% (equilibrium state) represent the law of mass action for the determined values of pK_a. While the experimental data agree with the mass action law for the monoamine, they diverge in the case of the polymer. Whether intra-molecular interactions or other effects are responsible for this deviation or other effects will be the subject of further investigations.

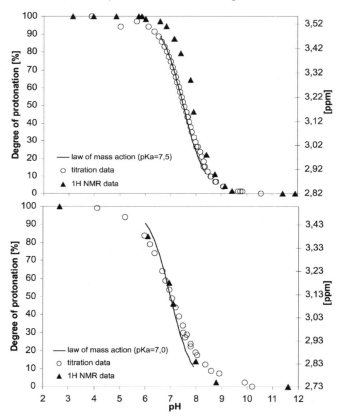

Fig. 1 Degree of protonation of PPP **27** (*bottom*) and the monoamine N[(CH$_2$CH$_2$O)$_2$ CH$_3$]$_3$ (*top*) in aqueous solutions as a function of pH; the chemical shifts δ(^{1}H NMR) of the CH$_2$ groups attached to the nitrogen atom are charted on the right *y*-axis

2.3
Further Approaches to PPP Polyelectrolytes

In addition to the research work presented so far, in the 1990s many other activities were directed towards the synthesis of rodlike polyelectrolytes. Some important examples will be referred to in the following. Regarding cationic polyelectrolytes in particular, Reynolds et al. described the synthesis of water-soluble PPPs such as **29** [28]. These polymers were analyzed with regard to their potential application as luminescent materials [29]. Similar polymers were reported by Swager et al. in 2000 [30]. Here, poly(*p*-phenylene ethynylenes) **30** were investigated as active components for chemosensors (Scheme 7).

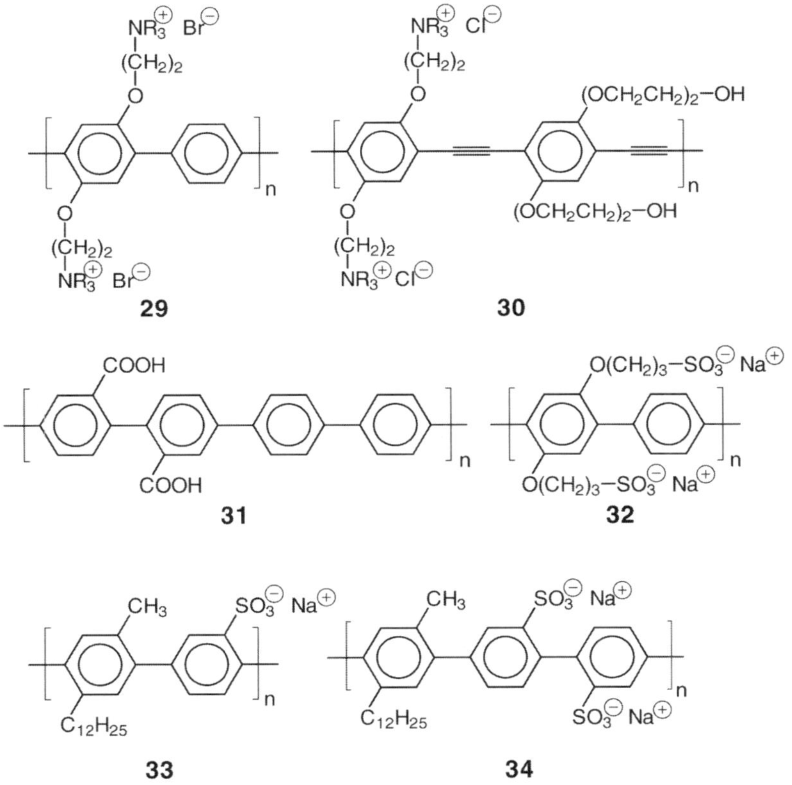

Scheme 7 Rodlike polyelectrolytes developed by other research groups

In the field of anionic polyelectrolytes, Wallow and Novak reported in 1991 on carboxylated PPPs such as **31** which proved to be soluble in aqueous bases [31–34]. Three years later, Reynolds et al. first synthesized sulfonated

PPPs such as **32** [35, 36]. Here, also, a direct in-aqua synthesis was applied. Simultaneously, Wegner et al. developed precursor syntheses of sulfonated PPPs such as **33** and **34** [37–42]. Due to their long aliphatic side chains, these latter polymers form well-defined cylindrical micelles in solution. They are of considerable current interest because they allow more to be learned about the association behavior of rodlike polyelectrolytes consisting of hydrophilic and hydrophobic sub-units. In this context, the studies represent an important supplement for the studies performed in our groups.

3
Polycations and Polycarbobetaines:
Narrowly Distributed Model Homo and Block Copolymers

Both polycarbobetaines and block copolymers containing a cationically charged block have been known for many years, but in the past only a few publications have dealt with their synthesis and properties in comparison with other types of polyelectrolytes. Increasing interest can be recognized in recent years, due to the close connection of polycarbobetaines to proteins and living matter [43], and due to the formation of novel supramolecular structures and nanoparticles with the block copolymers by self-assembly and interaction with oppositely charged compounds leading to several interesting applications [44, 45]. Polymeric betaines containing an equal number of permanent positively (quaternary ammonium group) and mostly negatively (sulfonate, sulfate, phosphate, carboxylate function) charged sites in strictly equal concentration on each monomer unit are often discussed [46]. The overwhelming majority of these polyzwitterions are polymeric sulfobetaines. Polycarbobetaines on the contrary are only a minority, probably because of the additional pH-dependence of the polymer properties.

The synthesis of these polycarbobetaines was carried out in two different ways–free radical polymerization of vinyl betaines or functionalization of reactive precursor polymers. The first route leads to polymers with 100% functionality, but often only with low molecular weight and at low rate. Furthermore, the molecular characterization is very difficult, because the polymer conformation in aqueous solution is very sensitive to pH and ionic strength, and in all chromatographic methods the polymeric zwitterions have a strong interaction with the columns. The second route has, as a first step, the advantage of a simple polymerization of functional monomers whose molecular characterization can usually be carried out without additional problems. Assuming 100% conversion during the functionalization reaction with different reagents a well characterized set of polymers with varied chemical structure but constant degree of polymerization is available. This may be very useful for the investigation of structure–property relationships.

The chemical structures of polycarbobetaines published up to now can be divided into three groups:

– Quaternary esters or amides of (meth)acrylic acid, in which the quaternary nitrogen is substituted by an alkylcarboxy group of different chain length. Examples are acetate [46, 47], propionate [48], butanoate [47, 49], and hexanoate [50] as those substituents.
– Quaternary polypyrrolidinium compounds derived from carboxyalkyl substituted diallylammonium salts, containing linear and branched alkylcarboxy groups with up to 10 methylene groups between N^+ and COO^-. Depending on the structure these polymers exhibit surfactant like properties and only limited solubility in water. They may form supramolecular structures. Also polyzwitterions containing two carboxylic groups per quaternary ammonium group are available from similar monomers leading to polymers with structural features common to both polyampholytes and conventional polyelectrolytes [52, 53].
– Polyzwitterions derived from polymeric heterocyclic vinyl compounds like *N*-vinylimidazole [54] and vinylpyridine [55, 56], containing short carboxyalkyl substituents at the quaternary nitrogen.

Furthermore, a polycarboxybetaine with a peptide main chain was synthesized from poly(methyl L-glutamate) [57]. Several zwitterionic monomers based on isobutylene with long alkyl spacers between the polymerizable group and the zwitterionic moiety exhibit surfactant properties. They do not undergo free radical homopolymerization, but are copolymerizable with other monomers [58].

Investigations of the properties of polycarboxybetaines include solubility measurement and viscometry at different pH and ionic strength in aqueous solution, and thermal analysis and X-ray studies in the bulk. Aggregation phenomena are only important when longer alkyl spacers are involved [47, 51]. Potentiometric titrations result in structure dependent "apparent" basicity constants [59]. Laser light scattering and ζ-potential measurement of a poly(methacrylamido) derivative have been discussed in terms of chain aggregation [48].

Amphiphilic block copolymers containing a cationic block may carry as the second block a hydrophilic or a hydrophobic one. Those containing a hydrophilic second block belong to the so called double-hydrophilic block copolymers (DHBC), which have recently been reviewed [60]. This promising class of polymers has potential applications in drug-carrier systems, gene therapy, desalination membranes, as switchable amphiphiles, and others [45, 60, 77].

Several synthetic strategies are used to produce block copolymers containing a cationic block. Because charged monomers are not polymerizable by ionic techniques, the synthesis of the required block copolymers can be carried out by free radical polymerization of ionic vinyl monomers using macroinitiators, by modifying one block of a block copolymer and by coupling of two readily synthesized blocks.

Cationic block copolymers containing a hydrophobic block as the second one are rather scarce. Examples are diblock copolymers of styrene and 4-vi-

nyl-*N*-ethylpyridinium bromide [61] and triblock copolymers of butyl methacrylate and a quaternary ester of methacrylic acid [62], both synthesized by
modification of a precursor. DHBC containing a cationic block in quaternized form are often derived from poly(2-vinylpyridine) or poly(4-vinylpyri-
dine), prepared via quaternization of a block copolymer from an ionic or
controlled radical polymerization [60]. Another strategy uses macroinitiators already containing the uncharged block of the desired DHBC. Using
poly(ethylene glycol) (PEG) monoazo-macroinitiators the cyclopolymeriza-
tion of diallyldimethylammonium chloride (DADMAC) results in triblock
copolymers with an ionic middle block and two PEG blocks because combination is the only termination reaction [63]. The polymers with blocks of
comparable length are powerful stabilizers in the emulsion polymerization
of styrene leading to monodisperse cationic latexes of high colloidal stability [63, 65]. Interaction with oppositely charged polyelectrolytes leads to na-
noscaled polyelectrolyte complex particles whose colloidal stability depends
on the length of the PEG chain [66]. PEG block copolymers containing poly-
soap blocks, representing a novel type of micellar polymer, can be synthesized in a similar way. By means of the same PEG macroinitiators the cy-
clopolymerization of *n*-alkyl-substituted quaternary diallylammonium compounds (*n*=8 or 12) having the properties of emulsifiers leads to water insoluble materials. However, water soluble polymers with high solubilization capacity can be obtained by copolymerization of the vinyl surfactants with
DADMAC [64, 65].

Recently diblock copolymers of PEG and ionic segments were prepared
by atom-transfer radical polymerization of methacrylic aminoester using a
monofunctionalized PEG macroinitiator and then subsequent quaternization. Like others [60] these polymers form so called polyion complex micelles by electrostatic interaction with oppositely charged molecules (e.g.
drugs, oligonucleotides), where the PEG block acts as a steric stabilizer [67].

3.1
Narrowly Distributed Cationic Polyelectrolytes and Polycarbobetaines

The synthesis of narrowly distributed cationic polyelectrolytes and polycarbobetaines has been carried out via controlled free radical polymerization of
reactive monomers. Using *N*-oxyls as terminators, polymers of vinylben-
zylchloride (VBC) [68–71] and 4-vinylpyridine (4-VP) [72, 73] with low
polydispersities (PD) are available. Usual PD-values are in the range of 1.2–
1.3 for VBC [68] and 1.3–1.5 for 4-VP [72]. These polymers serve as reactive
precursors in the synthesis of functionalized macromolecules. They can be
readily modified, leading to the desired ionically charged polymers with narrow molecular weight distribution. This strategy opens the possibility of
synthesizing polymers tailored towards the expected properties.

Quaternization of poly(4-VP) with alkyl halides and reaction of
poly(VBC) with various tertiary amines result in cationic polyelectrolytes
with a variety of charge density and of the hydrophilic–hydrophobic ratio of

the macromolecules but with the same degree of polymerization P_n [68, 72, 74, 75] (Scheme 8).

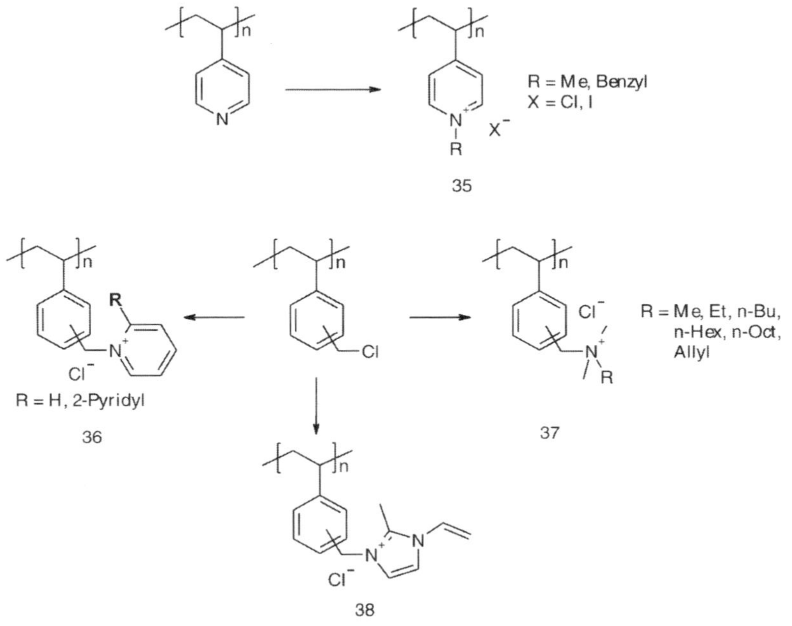

Scheme 8 Synthesis of cationic polyelectrolytes by functionalization of poly(4-VP) and poly(VBC)

They can serve as models for basic investigations of solution properties and interaction with oppositely charged matter leading to new structure–property relationships. A further advantage is the low PD supporting the precision of any measurement.

Typical examples are:

- Comparing polymers **37** (R=n-Bu), **36** (R=H) and **36** (R=2-pyridyl) the increasing stiffness of the chains due to the more bulky substituents leads to an increase of the exponent in the Mark–Houwink equation from 0.65 over 0.77 to 1.00. Variation of the ionic strength as well as of the kind of counter ion has a remarkable influence on α. With polymer **37** (R=Me) the coil expansion during decreasing ionic strength results in higher α-values. Decreasing counter ion radius again causes higher α-values [74, 75].
- Experimental and calculated data of the partial molar volume of polymers **36** (R=H), **37** (R=allyl), **37** (R=n-Bu, n-Oct) are in good agreement. In contrast to investigations with polyanions no influence of the molecular weight is detectable [75].

- Electrochemical measurement of polymers **36** (R=H), **37** (R=allyl) and **37** (R=Me, *n*-Bu, *n*-Oct) contribute to theoretical models of polyelectrolytes. A pronounced influence of concentration and chain length on the polyion–counterion interaction was detected; this has not been evident from existing models. Changes in the ionic interactions are primarily responsible for the dependence of the equivalent conductivity on the molecular weight below the overlap concentration. With polymers of equal P_n a clear influence of the chemical structure is detectable [76].
- The electrophoretic mobility as measured by capillary electrophoresis (CE) depends on the chemical structure of the polyelectrolytes. Comparing **37** (R=Me) and **37** (R=*n*-Bu) the latter has a distinctly lower mobility, probably caused by some hydrophobic interaction of the butyl moiety leading to shielding of the charged nitrogen [78]. After sufficient calibration a separation following the molecular weight and a determination of M_w and M_n for a given polyelectrolyte is possible [79].
- Polymers **36** (R=H) and **37** (R=Me, *n*-Bu) have been successfully applied as displacers in ion-exchange displacement chromatography (77).

Poly(4-VP) and poly(VBC) are useful precursors for the synthesis of polycarboxybetaines also. Alkylation of poly(4-VP) with bromocarboxylic acid esters and quaternization of poly(VBC) with amino acid esters, both followed by saponification of the ester moiety, led to the narrowly distributed polymers [64, 80]. A series of polycarboxybetaines of the pyridiniocarboxylate and of the ammonioacetate type with very high degree of functionalization was prepared [80] thereby varying the alkyl spacer length between both charges, the length of an additional alkyl chain at the α-carbon, and different substitution at and hybridization of the quaternary nitrogen (Scheme 9).

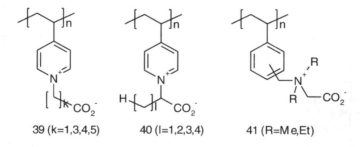

Scheme 9 Polycarboxybetaines synthesized by functionalization of poly(4-VP) and poly(VBC)

The pH-dependent solution properties in aqueous systems and the interaction of the different charges of the molecules were measured by CE and charge titration at different pH. CE is an especially powerful tool for characterization of the behavior of polycarboxybetaines in solution [78, 79]. A clear influence of the chemical structure was observed [64, 78–80]. An increase of the mostly intramolecular interaction of polymers **39** and **40** as revealed by lower mobility in CE and lower amount of titratable cation during charge titration was detected with decreasing length of both the spacer between the ionic moieties and the alkyl substituent at α-C. This result was supported by FTIR measurement in bulk. Comparing polymers **41**, increasing shielding of the quaternized nitrogen (**41**, R=Et) results in a lower charge interaction. In general, with any structural variation an increase of charge interaction with increasing pH was observed.

For equal distance of the charged sites, for example with **39** (k =1) and **41** (R=Me), the electrophoretic mobility of the polymer containing N with sp²-hybridization is much lower, because of delocalization of the positive charge in the aromatic ring.

Typical examples of CE are given in Figs. 2 and 3.

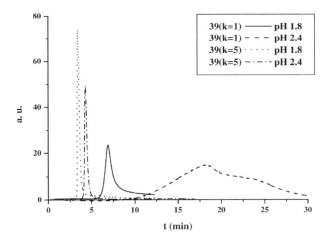

Fig. 2 Electrophoretic mobility of polymers 39(k=1) and 39(k=5) at pH 1.8 and 2.4

Complexing betaines **39** and **40** with strong polyanions such as polystyrenesulfonate sodium salt reduces intramolecular interaction and acid–base titration becomes possible [85]. In pure poly(4-VP) based polycarbobetaines only 10–15% of the carboxylic groups are titratable [64].

At very low pH-values an increase of the molecular weight of polycarboxybetaines with more limited interaction of charged groups was observed by ultracentrifugation in the sedimentation equilibrium, which is attributed to the formation of H-bridges by the carboxylic groups [81].

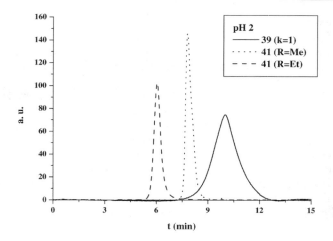

Fig. 3 Electrophoretic mobility of polymers 39(k=1), 41(R=Me) and 41(R=Et)

3.2
Ionically Charged Block Copolymers

The N-oxyl-mediated controlled free radical polymerization of 4-VP and VBC can easily be continued with styrene (ST) as second monomer to synthesize the corresponding block copolymers [68, 72]. In principle all the reactions leading to the structures in Schemes 8 and 9 can be applied to these reactive precursors, leading to a wide variety of block copolymers containing a strongly hydrophobic polystyrene and a hydrophilic cationic or betaine block with adjustable block length and block length ratio [68, 72, 81, 82]. Examples are given in Scheme 10.

Polymers **42** are micellarly soluble in water and in organic solvents. As expected, the aggregation number and the sedimentation coefficient and the radius of gyration increase with increasing length of the hydrophobic block, while keeping the hydrophilic one constant [64, 68, 82]. The micellization process is not in thermodynamic equilibrium at room temperature, resulting in so-called frozen micelles. A frozen equilibrium can be established by tempering above T_G of poly(ST) and rapid cooling [64]. Viscosity measurement of **42** reveals intermolecular interactions of the charged parts to be active in pure water [64].

Polymer **42** (R=Me) is a powerful stabilizer in emulsion polymerization of ST. The particle size increases with increasing length of the hydrophobic block. Due to very efficient electrosteric stabilization the resulting dispersions have outstanding colloid stability even at high salt concentration [64, 75].

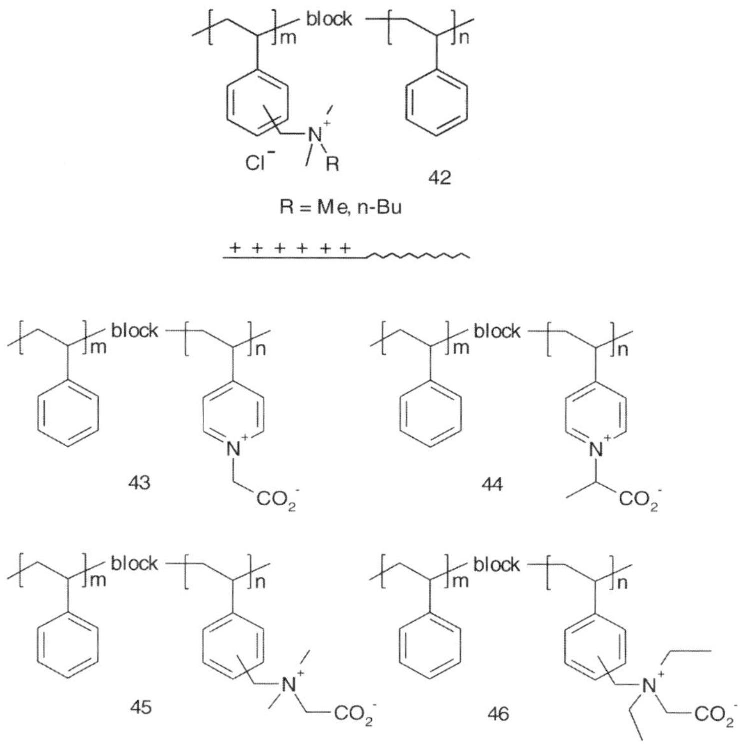

Scheme 10 Structure of block copolymers containing a cationic or betainic block

The complexation of **42** (R=Me) with poly(acrylic acid) (PAA) results in a mesomorphously ordered material consisting of hexagonally-ordered ion-rich cylindrical rods containing the PAA embedded in a polystyrene matrix. This complex represents a new type of a polymeric hybrid material with a supramolecular ordered nanostructure [83].

The betaine-containing block copolymers **43**, **44** and **46** show a high tendency to aggregate in aqueous solution. Only **45** is soluble without limit. Investigations by ultracentrifugation show an increase of the sedimentation and diffusion coefficient and of the diameter of the aggregates with increasing ratio of the block length of ST vs. carbobetaine [81].

A novel route to DHBC is outlined in Scheme 11. Reaction of TEMPO-terminated poly(VBC) with PEG-monomethyl ether results in reactive block copolymers **47**, which can be converted readily to polymers **48** and **49** containing a cationic or a betainic block besides a PEG block [84].

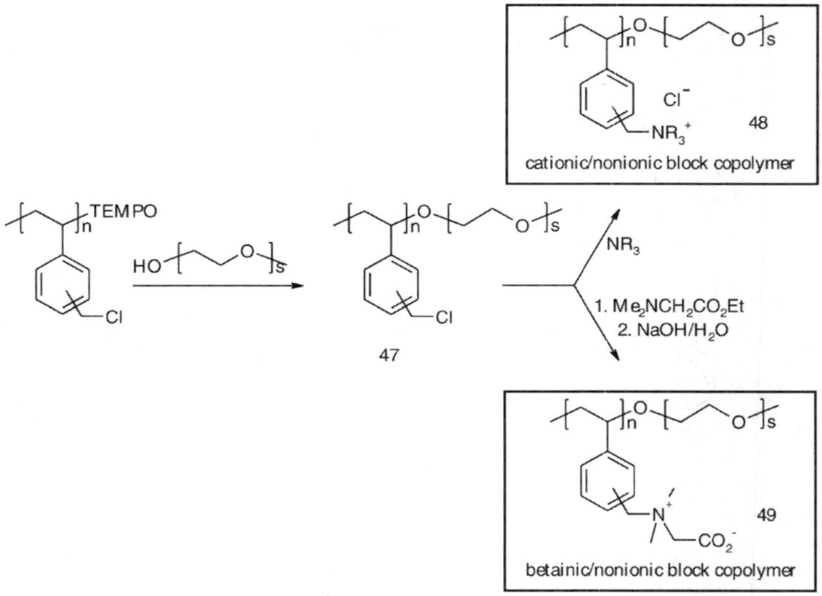

Scheme 11 A novel route to double hydrophilic block copolymers

4
Nonlinear Polyelectrolyte Topologies

Branched polyelectrolytes (PEs) have become of special interest because of their industrial importance and scientifically interesting properties. Poly (ethyleneimine), which is important in various industrial applications, provides an excellent example–branched and linear polyelectrolytes have quite different properties due to both the different topographies and structures [43, 86, 87]. As an another practical point, branched polyelectrolytes can act as predecessors or fragments of polyelectrolyte gels. Variation in the degree of branching (DB) leads to a continuous change in the properties of branched macromolecules from linear chains to soft nanoparticles with highly compact structures, due to their branched architectures. A variety of theoretical approaches have been reported in the investigations of branched polyelectrolytes [88–93]. However, correlation of the topology and properties of branched polyelectrolytes has not been studied very much experimentally, because of difficulties in the synthesis of well-defined branched polymers with ionic or ionizable groups. One challenge in this field is, therefore, to produce randomly or regularly branched polyelectrolytes which are suitable for various applications as well as for quantitative analysis of the relation between the properties and the architectures. The material properties of branched polymers depend not only on the molecular weight (MW) and DB, but also on the type of branching. It is, therefore, desirable to establish precise synthetic methods for various topologies of branched polymers (random, comb, and star).

This chapter will focus on recent advances in the synthesis of novel branched polyelectrolytes using controlled polymerization methods. For the synthesis of well-defined polymers, living polymerization techniques are traditionally employed in which the polymerizations proceed in the absence of irreversible chain transfer and chain termination. However, the application of these techniques for the synthesis of branched polyelectrolytes was limited, as most of the systems are less tolerant of functional groups. Recent development in controlled radical polymerization (CRP) methods has provided another methodology to synthesize branched polyelectrolytes. The systems include atom transfer radical polymerization (ATRP) [94–97], nitroxide-mediated radical polymerization [98], and reversible addition–fragmentation chain transfer polymerization (RAFT) [99]. All systems are based on establishing a rapid dynamic equilibration between a minute amount of growing free radicals and a large majority of dormant species, and are more tolerant of functional groups and impurities. However, with the exception of RAFT, no controlled anionic, cationic, or radical polymerization method is able to polymerize acidic monomers, like acrylic acid (AA) or styrenesulfonic acid. Thus, typically protected monomers, like *tert*-butyl acrylate (tBuA), have been employed, followed by a polymer-analogous reaction, e.g. hydrolysis of the protecting ester groups. Similarly, cationic polymers containing quaternized amine functionalities, are derived from non-quaternized monomers, e.g. vinyl pyridine or 2-dimethylaminoethyl methacrylate.

4.1
Star-Shaped Polyelectrolytes

There are two basic synthetic routes for star polymers (Scheme 12)–the "core first" method (polymerization from multifunctional initiators or microgels) and the "arm first" method (where growing polymer chain ends are reacted with a multifunctional terminating agent or a divinyl compound). Whereas the use of multifunctional initiators leads to stars with a well-known (but often low) number of arms, the use of microgels or divinyl compounds leads to a rather broad arm number distribution, where the average arm number can be quite high.

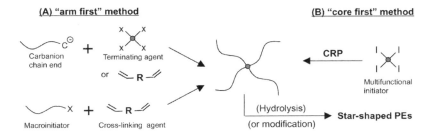

Scheme 12 Synthesis of star-shaped polyelectrolytes by the **a** "arm first" and **b** "core first" methods

Living anionic polymerization with the "arm first" method has been applied for the synthesis of star-shaped polyelectrolytes. Linear polymers with narrow molecular weight distribution (MWD) and well-defined MW are achieved by anionic living polymerization, and the star-shaped structures are formed by reaction of carbanion chain ends with a multifunctional terminating agent or divinyl compound. Several star polymers consisting of poly(2-vinylpyridine) (P2VP), a weak cationic polyelectrolyte, have been synthesized by living anionic polymerization [100–104]. Star-shaped poly(styrenesulfonate)s, strong anionic polyelectrolytes, have been also reported [105, 106].

Star-shaped poly(acrylic acid) (PAA) and poly(methacrylic acid) (PMAA) have been synthesized by anionic living polymerization of the protected monomers. However, the controlled synthesis of well-defined star polymers of (meth)acrylates is still a challenge, because of the reduced reactivity towards multifunctional terminating agents at the low temperatures which typically must be used in order to avoid side reactions. The synthesis of Pt-BuA star polymers by anionic polymerization was conducted from living multifunctional microgel cores prepared by reaction of divinylbenzene with lithium naphthalenide or short living polystyrene chains [107]. PtBuMA star polymers were prepared via living anionic polymerization using two different terminating agents, 1,3,5-trisbromomethylbenzene and octa[(3-iodopropyl)]silsesquioxane, in order to further react them to form model polyelectrolyte networks [108]. Subsequent cleavage of the *tert*-butyl ester moieties by acidic hydrolysis gave access to photo-crosslinkable, monodisperse polyelectrolyte star polymers. A series of star polymers consisting of PtBuA arms and an ethylene glycol dimethacrylate (EGDMA) microgel core were synthesized using anionic polymerization [109]. The effect of the various reaction parameters on the average number of arms and solution properties, such as radius of gyration and intrinsic viscosity, of the star-shaped PtBuAs was investigated using GPC coupled with viscosity and multi-angle laser light-scattering detectors. Star-shaped PAAs were obtained after hydrolysis of the *tert*-butyl esters [110].

The formation of PAA star polymers using the "core first" method has been demonstrated in the ATRP process by use of multifunctional initiators [111, 112]. In this method, the number of arms in the star polymer can be determined by the number of initiating sites on the initiator. Star-shaped PtBuA was prepared by the "arm first" method via ATRP, using divinylbenzene, 1,4-butanediol diacrylate, and EGDMA as coupling reagents [113].

In spite of the synthetic success in forming polyelectrolyte stars, little research has been published on their solution properties. PAA stars obtained from hydrolysis of PtBuA stars (made by the core-first method and coupling with EGDMA [109]) having an arm number distribution were investigated by GPC coupled with viscosity and light-scattering detectors [110]. The results were compared to those obtained with the non-ionic precursors. Contraction factors were obtained by comparing the intrinsic viscosities, $g' = [\eta]_{star}/[\eta]_{linear}$ and the radii of gyration, $g = R^2{}_{g,star}/R^2{}_{g,linear}$ of star-shaped and linear PAANa (measured in water with 0.1 mol L^{-1} NaNO$_3$) and PtBuA

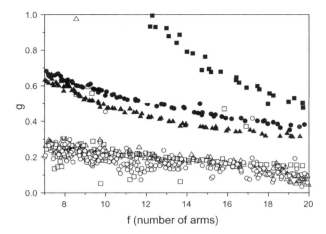

Fig. 4 Contraction factors, $g=R^2\hat{g}_{,star}/R^2\hat{g}_{,linear}$, as a function of the number of arms, f, for three different PAA stars (*closed symbols*) and non-hydrolyzed PtBuA precursor stars (*open symbols*) [110]

(the non-ionic precursors, measured in THF). As is shown in Fig. 4 for the g values, the contraction of polyelectrolyte stars is much smaller than that of non-ionic stars. This was attributed to the stronger repulsion of the ionic charges on the polyelectrolyte stars, especially in the center of the stars. Thus addition of further arms leads to an increase in charge density which is compensated by arm stretching and an increase of R_g, which is larger than for non-ionic polymers.

4.2
Randomly Branched (Arborescent) Polyelectrolytes

Randomly branched polyelectrolytes contain several branching points which are distributed irregularly in the polymer chains. If a certain degree of branching (DB, defined as the fraction of branchpoints and endgroups) is exceeded, arborescent topologies are formed, carrying branches on branches. These are frequently called hyperbranched polymers. It is expected that the DB has a significant effect on the solution properties of polyelectrolytes.

The recent discovery of self-condensing vinyl polymerization (SCVP) made it possible to use vinyl monomers for a convenient, one-pot synthesis of hyperbranched vinyl polymers with DB≤0.5. The initiator–monomers ("inimers") used are of the general structure AB*, where the double bond is designated A and B* is a group capable of being activated to initiate the polymerization of vinyl groups. Cationic [114], anionic [115], group transfer [116], controlled radical [117–121], and ring-opening mechanisms [122] have been used. By copolymerizing AB* inimers with conventional monomers, this technique was extended to self-condensing vinyl copolymerization (SCVCP), leading to highly branched copolymers with DB controlled by the comonomer ratio [123–126]. Depending on the chemical nature of the comonomer, ionic or ionizable groups can be incorporated in the branched polymer, leading to highly branched polyelectrolytes.

Highly branched PtBuMA was synthesized by SCVCP of tBuMA with an inimer having a methacrylate group (A) and a silylketene acetal group (B*), capable of initiating group-transfer polymerization [125]. Characterization using multidetector SEC indicated that the corresponding intrinsic viscosity of the resulting copolymers was lower than that of linear PtBuA, suggesting a branched structure. Acid-catalyzed hydrolysis of the *tert*-butyl groups and neutralization with NaOH produced a water-soluble, highly branched PMAA sodium salt.

The synthesis of randomly branched PAA was conducted by SCVCP of tBuA with an inimer having an acrylate (A) and an α-bromopropionate group (B*), capable of initiating ATRP, followed by hydrolysis of the *tert*-butyl groups (Scheme 13) [127]. Characterization of the branched PtBuAs was

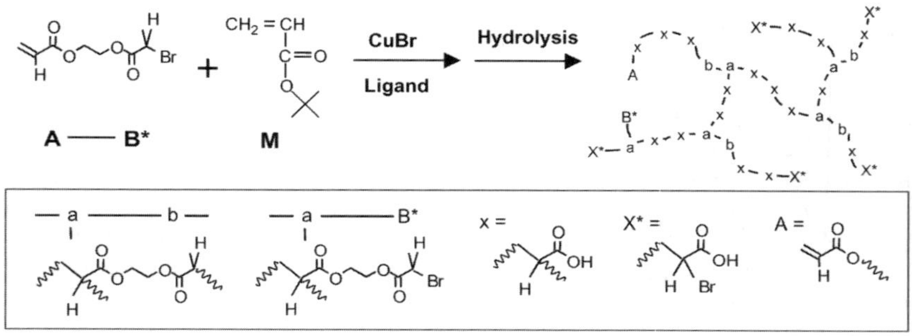

Scheme 13 General route to branched poly(acrylic acid) (PAA) via self-condensing vinyl copolymerization (SCVCP) of an inimer, AB* and a monomer, M, followed by hydrolysis

conducted by GPC, GPC/viscosity, GPC/multiangle laser light scattering, and NMR analysis, demonstrating that DB, the composition, MW, and MWD can be adjusted by appropriate choice of the catalyst system, the comonomer composition in the feed, and the polymerization conditions. The viscosities of the branched PtBuAs in THF are significantly lower than those of linear PtBuA and decrease with increasing degree of branching which, in turn, is determined by the comonomer feed ratio, $\gamma=[\text{tBuA}]_0/[\text{inimer}]_0$. The Mark–Houwink exponents are significantly lower ($\alpha=0.38$–0.47) compared to that for linear PtBuA ($\alpha=0.80$). Even for $\gamma=100$ (corresponding to only 1% inimer), the α value is only 50–60% of the value of linear PtBuA. The contraction factors decrease with increasing MW, indicating a highly compact structure in solution. Selective hydrolysis of the branched PtBuAs to PAA was confirmed by ^{1}H NMR, FTIR, aqueous-phase GPC measurement, and elemental analysis. The results confirm that the branched structure is intact during the hydrolysis. The water solubility of the branched PAAs decreases with increasing DB and decreasing pH. Potentiometric titration curves (Fig. 5) suggest that the apparent pK_a values (taken as the pH at 50% ionization) of the branched PAAs with a comonomer ratio $\gamma=10$ and 100 are comparable to the corresponding value for linear PAA homopolymer of $pK_{a,app}\cong5.8$ [128]. The branched PAA at $\gamma=2.5$ become insoluble at pH$\leq$4.7. For the PAAs with a higher degree of branching ($\gamma\leq1.5$), solubility is obtained only at pH>8, suggesting that these polymers are soluble in water only at higher degree of ionization. However, for low comonomer ratios it has to be taken into account that a large fraction of the polymer (i.e. the linking inimer groups) is, in fact, non-ionic in nature. Aqueous-phase GPC and dynamic light scattering confirm the compact structure of the randomly branched PAAs. Comparison of the hydrodynamic radius of branched and linear PAAs (obtained from dynamic light scattering, Fig. 6) as a function of M_w suggested that highly compact structures of the polyelectrolytes can be obtained due to their branched architectures. Studies at different pH indicate that marked stretching of the branched chains takes place when going from a virtually uncharged to a highly charged stage.

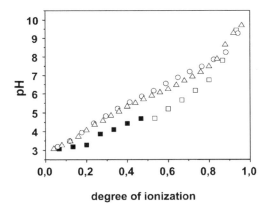

degree of ionization

Fig. 5 Potentiometric titration curves for branched PAAs with comonomer ratios $\gamma=100$ (*circles*), 10 (*triangles*), 2.5 (*squares*) in aqueous solutions. The *filled squares* indicate the region where the branched PAA at $\gamma=2.5$ was insoluble in aqueous solution. (Reproduced with permission from Ref. [127]. Copyright 2002 American Chemical Society.)

Another approach involved the SCVP of a "macroinimer", which is a heterotelechelic PtBuA possessing both initiating and polymerizable moieties, via ATRP [129]. GPC/viscosity measurements indicated that the intrinsic viscosity of the branched polymer is less than 40% that of the linear one at highest MW area. A significantly lower value for the Mark–Houwink exponent ($\alpha=0.47$ compared to $\alpha=0.80$ for linear PtBuA) was also observed, indicating the compact nature of the branched macromolecules.

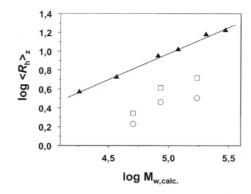

Fig. 6 Dependence of z-average hydrodynamic radius, R_h, on weight-average molecular weight, M_w of the branched PAAs: R_h was measured at pH=3 (*circles*) and pH=10 (*squares*). Data points for linear PAAs (*triangles*) used as a reference were taken from Reith et al. [154]. (Reproduced with permission from Ref. [127]. Copyright 2002 American Chemical Society.)

4.3
Branched Polyelectrolytes Grafted on Surfaces

Polyelectrolyte chains attached to planar and spherical surfaces have recently attracted much interest as academic model systems and as candidates for various industrial applications. In most studies, the grafted polyelectrolyte is generated by adsorption of end-functionalized polymers or block copolymers on the surfaces [130, 131]. These systems, however, have a limited grafting density as further grafting is hindered by the polymer chains already adsorbed on the surface. Surface-grafted polyelectrolytes have been also synthesized by "grafting from" techniques via conventional radical polymerization [132] or photoemulsion polymerization [133–136], achieving higher grafting density. However, there is poor control over chain length and chain end functionality.

CRP from surfaces [130, 137] allows better control over MW and MWD of target polymer. Only few approaches have been reported on the CRP grafting of polyelectrolytes from planar and spherical surfaces (Scheme 14a). A cross-linked polystyrene latex functionalized with alkyl bromide group was used as an substrate for the ATRP of 2-(methacryloyloxy)ethyl trimethylammonium chloride [138]. In order to amplify initiators patterned on films of gold ATRP was applied for grafting of poly(2-(dimethylamino)ethyl methacrylate) and PtBuMA grafted on a planar gold surface [139]. ATRP was also used for grafting a polystyrene-b-PtBuA block copolymer from silicon, followed by conversion to polystyrene-b-PAA [140].

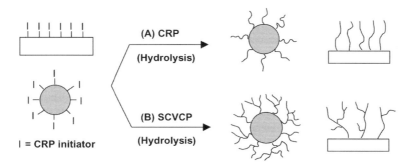

Scheme 14 Surface-grafted linear and branched polyelectrolytes (or their precursors) prepared by **a** CRP and **b** SCVCP from flat surface and spherical particle

The surface chemistry and interfacial properties of highly branched polymers have also become fields of growing interest [141–148]. Recently, surface-initiated SCVCP has been applied as a new method for the synthesis of highly branched polymers grafted from planar and spherical surfaces (Scheme 14b). Branched PtBuA–silica hybrid nanoparticles were synthesized by surface-initiated SCVCP of an acrylic AB* inimer with tBuA from silica nanoparticles functionalized with monolayers of ATRP initiators [149]. Well-defined polymer chains were grown from the surface to yield hybrid nanoparticles comprising silica cores and highly branched PtBuA shells, as confirmed by elemental analysis and FTIR measurements. Hydrolysis of the ester functionality created highly branched PAA–silica hybrid nanoparticles. The chemical composition and the architectures of the branched PAAs grafted on the silica nanoparticles were controlled by the composition of the feed during the SCVCP. Surface-initiated SCVCP was also applied for the synthesis of branched PtBuA grafted on silicon [150]. These methodologies can be applied to a wide range of inorganic materials for surface-initiated SCVCP to allow the preparation of new 2D and 3D branched polyelectrolyte/inorganic hybrid materials.

There are several accounts of the preparation of branched polyelectrolytes grafted on surfaces which did not rely on CRP. A highly branched PAA film attached to a self-assembled monolayer of mercaptoundecanoic acid on flat gold surface has been prepared using a series of repeated steps by the "grafting to" technique [151, 152]. Branched poly(ethyleneimine) brushes were synthesized by ring-opening polymerization of aziridine from amine-functionalized surfaces [153]. Such branched polyelectrolytes grafted on surfaces may have interesting properties due to their architecture and confined structure.

5
Well Defined Amphipolar Poly(Acrylic Acid-*Graft*-Styrene) Copolyelectrolytes

The high potential of amphipolar copolymers as stabilizers for particulate systems has been shown in recent years especially as far as emulsions and dispersions are concerned [155, 156]. One promising area of application of copolymers is the stabilization of pigments–by using amphipolar copolyelectrolytes both electrostatic and steric effects can be combined [157, 158] in the so-called electrosteric stabilization [159, 160]. In order to obtain optimal stabilization of the particulate system one can take advantage of the variable macromolecular architecture as given in random, block or graft copolymers: Synergistic effects can be achieved by incorporating different structural elements of ionic, hydrophilic and hydrophobic nature [161]. Theoretical [162, 163] and experimental work [164–166] has shown that amphipolar graft copolymers provide a better stabilization performance for pigment dispersions as compared to random or block copolymers. Furthermore, graft polymers have been of major interest for other applications in polymer science [167, 168].

In order to explore the potential of amphipolar graft copolymers as polymeric stabilizers and to investigate the relationship between the copolymer structure and interaction of the copolymer with the particle surface, a series of well defined styrene (S)/acrylic acid (AA) model graft copolymers was synthesized [166]. There are basically three routes for preparing amphipolar graft copolymers, i.e. "grafting-from" [169, 170] or "grafting onto" [171, 172], and via macromonomers [165, 173]. Whereas the first two techniques are very limited in controlling the constitution of the graft copolymer, access to regularly structured graft copolyelectrolytes with an ionic poly(acrylic acid) (PAA) backbone and hydrophobic polystyrene (PS) grafts is given by the macromonomer technique (cf. [174, 175]). The length and the length distribution of the PS grafts is predetermined and easily adjusted by the anionic polymerization technique of the macromonomer synthesis; the backbone length may be controlled by the use of a chain transfer reagent in the copolymerization of the acrylic monomer with PS macromonomer, e.g. benzyl mercaptan, which has been shown to be an effective transfer agent in radical polymerization [176, 177a].

The preferred strategy for synthesis of amphipolar graft copolyelectrolytes consisting of a PAA backbone of controlled average degree of polymerization and adjusted PS graft density of known and relatively narrow length distribution is to copolymerize a methacrylate functionalized PS macromonomer with an acrylate comonomer followed by a polymer analogous reaction, i.e. saponification of the acrylate. The direct synthesis of poly(acrylic acid–*graft*-styrene) (PAA-*g*-PS) by copolymerization of PS macromonomer with (meth)acrylic acid (AA) is also possible [173, 177b], but purification of the crude polyelectrolyte and analysis of the graft copolymer constitution is problematic [178b]–the amphipolar character of the polyelectrolyte makes the work-up more difficult and causes relatively low yields; furthermore, compositional analysis by ^1H NMR spectroscopy requires esterification of the AA carboxylic group by polymer-analogous reaction (e.g. with diazomethane) since the spectrum of PAA–g-PS does not exhibit suitably separated signals.

The reaction scheme for graft copolyelectrolyte synthesis by free radical copolymerization according to the macromonomer technique is shown in Scheme 15. Besides the aspect of how to control the constitution of the graft copolyelectrolyte, suitable characterization techniques for unequivocal proof of the attained copolymer structure will also be elucidated. The synthesis, characterization and properties of the inversely structured poly(acrylic acid)–g-polystyrene graft copolymers it are covered in another article in this volume [178].

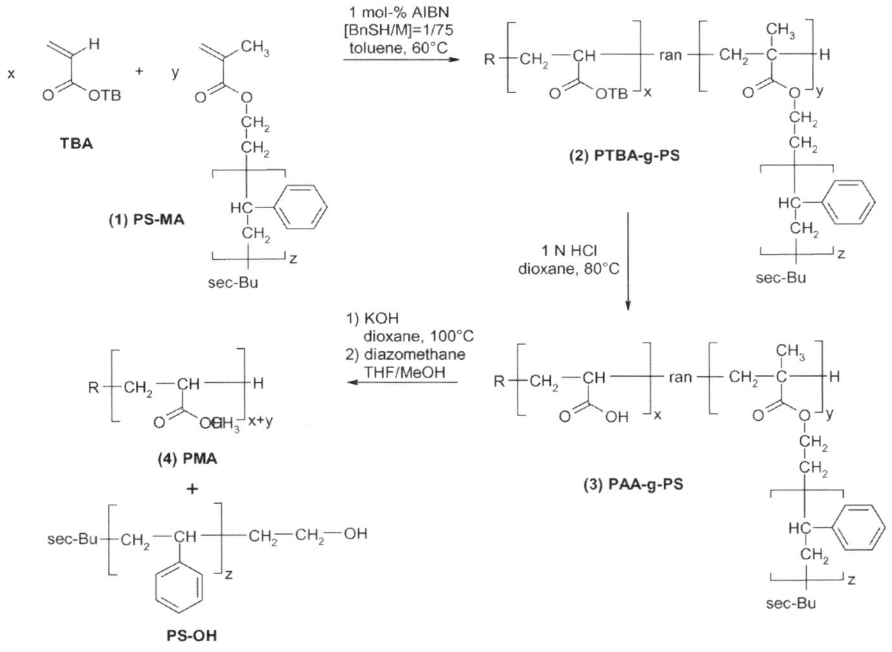

Scheme 15 Reaction scheme for the synthesis of poly(acrylic acid–*graft*-styrene) copolymers (PAA–g-PS) according the macromonomer technique

5.1
Control of Poly(Acrylic Acid) Backbone Length by a Chain Transfer Reagent

The possibility of controlling backbone length in the free radical copolymerization of TBA monomer with PS-MA macromonomer has been studied by radical polymerization of TBA in the presence of various amounts of benzyl mercaptan (BnSH) as chain-transfer agent. Under the assumption that each polymer chain is started by a benzyl mercaptan radical, as described in the literature [179], the degree of polymerization is calculated from the ratio of the intensities of the ^{1}H NMR signals of the aromatic benzyl mercapto end group protons (7.1–7.4 ppm) and the aliphatic protons of the *tert*-butyl group (1.3–1.6 ppm). A representative ^{1}H NMR spectrum of a synthesized poly(*tert*-butyl acrylate) (PTBA), and peak assignment of the molecular structure, are shown in Fig. 7a. The variation of the molecular weight as determined by ^{1}H NMR or SEC of the poly(*tert*-butyl acrylate) (PTBA) synthesized with different benzyl mercaptan/*tert*-butyl acrylate ratios [BnSH/TBA] is shown in Fig. 7b.

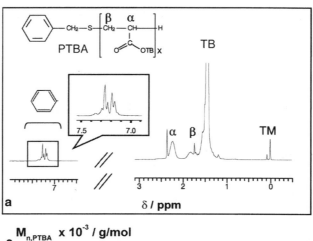

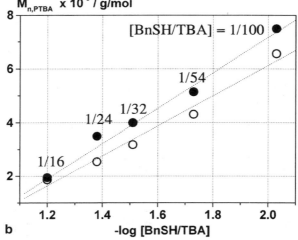

Fig. 7 (a) ^{1}H NMR spectra of poly(*tert*-butyl acrylate) (PTBA) synthesized by free radical polymerization of *tert*-butyl acrylate (TBA) in the presence of benzyl mercaptan (BnSH). (b) Dependence of the number average molecular weight ($M_{n,PTBA}$) of the PTBA on the BnSH/TBA ratio as obtained from ^{1}H NMR end group analysis (*filled circles*) and SEC (*open circles*) (the actual BnSH/TBA molar ratio is indicated)

The data represented in Fig. 7b illustrate that the degree of polymeriza-tion, i.e. the molecular weight, can be readily adjusted in the M_n-range be-tween about 2000 and 8000 g mol^{-1} by use of the appropriate chain transfer agent/monomer ratio; the deviation between the M_n-values obtained by ^{1}H NMR end group analysis and SEC are due to the PS calibration in the SEC; the latter figures reflect the somewhat smaller hydrodynamic volume of PTBA as compared to PS in THF solution. For the synthesis of the graft copolymers by the macromonomer technique a benzyl mercaptan/*tert*-butyl acrylate ratio [BnSH/TBA]=1/75 was used in order to reach a number average degree of polymerization about 50, corresponding to a poly(acrylic acid) (PAA) backbone number average molecular weight of about $M_{n,PAA}$ =3500 g mol^{-1}.

5.2
Synthesis and Characterization of Oligomeric Polystyrene Macromonomers PS-MA

The synthesis and purification of polystyrene methacryloyl macromonomers (PS-MA) in the molecular weight range M_n=1000–2000 g mol^{-1} by living an-ionic polymerization of styrene (S), termination with ethylene oxide (EO), and subsequent reaction with methacrylic chloride has already been de-scribed in detail elsewhere [180] (see also Scheme 16). In this context it has to be emphasized that the hydroxyethyl-terminated PS-MA macromonomer precursor (PS-OH) as obtained after purification of the crude PS-OH by sili-ca column chromatography (cyclohexane/dichloromethane 1/1 v/v) and as charged in the PS-MA synthesis still contains up to about 15 wt-% of non-functionalized polystyrene (PS-H). This PS-H impurity of the PS-MA macro-monomer does not interfere with the PS-MA synthesis and the subsequent TBA/PS-MA copolymerization and is easily and conveniently removed from the resulting PTBA–*g*-PS graft copolymer (see below).

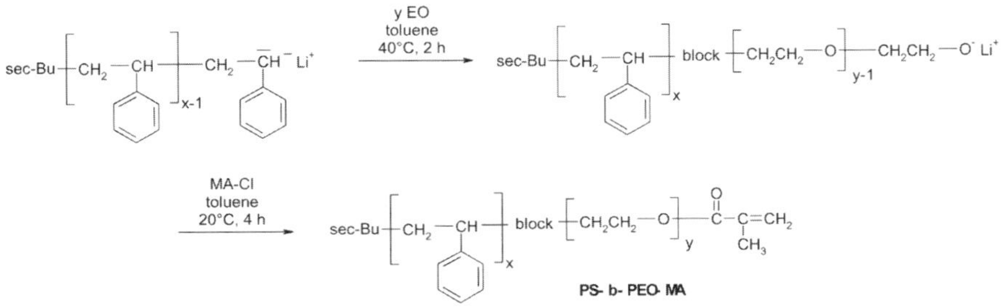

Scheme 16 Reaction scheme for polystyrene (PS) macromonomer synthesis

One interesting aspect of this PS-MA synthesis is the possibility of mod-ifying the amphipolarity of these macromonomers in the end capping reac-tion of the polystyryl anion with ethylene oxide (EO), as shown schemati-cally in Scheme 16. Contrary to the literature report [181] that only one EO unit is added to the carbanion chain end in benzene solution under the re-action conditions used (40 °C), we have found that a higher conversion of the EO present in the reaction mixture (see reaction Scheme 16) takes place within hours, resulting in an oligomerization of EO at the poly(styryl)lithium chain end. The EO conversion depends on the selected temperature and reaction time, and four/five ethyleneoxy units could easily be added to the carbanionic chain end. EO oligomerization during the functionalization of poly(styryl)lithium has also recently been described in the literature [182].

If such EO oligomerization occurs, an amphipolar poly(styrene-*block*-ethylene oxide) methacryloyl macromonomer (PS-*b*-PEO-MA) is obtained after further end group functionalization with methacryloyl chloride. As it is evident from the molecular structure (see Scheme 16) of the possible amphipolar PS-*b*-PEO-MA macromonomer, such hydrophilically modified macromonomers offer the opportunity to further tune the amphipolarity of the graft copolyelectrolytes; this will be investigated in further studies.

5.3
Synthesis of Amphipolar Poly(Acrylic Acid-*Graft*-Styrene) Graft Copolymers (PAA-*g*-PS)

With the above polystyrene macromonomers (PS-MA) of $M_{n,PS}$=1000 or 2000 g mol^{-1} and for the established benzyl mercaptan/*tert*-butyl acrylate ratio [BnSH/TBA]=1/75 first a series of PTBA-*g*-PS graft copolymers is synthesized by variation of the comonomer ratio in the macromonomer technique. PAA-*g*-PS copolymers are obtained by polymer analogous ester cleavage of the tert-butyl group of poly (*tert*-butyl acrylate-*graft*-styrene) (PTBA-*g*-PS) under acidic conditions; the selective *tert*-butyl ester cleavage is achieved by reacting PTBA-*g*-PS with hydrochloric acid (HCl) in dioxane solution at 80 °C for 4 h (10 mL dioxane+1 mL 6 mol L^{-1} HCl per g copolymer) [177a, 183]. The resulting PAA-*g*-PS is purified by re-precipitation from alkaline aqueous solution (~10 wt-%) with 6 mol L^{-1} HCl and obtained in yields of about 65 to 80%. The complete reaction scheme is shown in Scheme 15.

The composition of the PAA-*g*-PS graft copolymer reaction product and its purification, especially as far as the removal of unreacted PS-MA macromonomer by silica column chromatography is concerned, and the successful selective cleavage of the *tert*-butyl ester under acidic conditions to render the graft copolyelectrolyte PAA-*g*-PS were analyzed by ^{1}H NMR spectroscopy and SEC. Figure 8a shows the SEC curves of the polystyrene macromonomer (PS-MA), the crude poly (*tert*-butyl acrylate-*graft*-styrene) (PTBA-*g*-PS) and the PTBA-*g*-PS; the polymethylacrylate (PMA) originates from esterification of the poly (acrylic acid) (PAA) obtained after complete saponification of the graft copolymer and represents the backbone. The ^{1}H NMR spectra of PSMA, PTBA-*g*-PS and of the final reaction product PAA-*g*-PS are shown in Fig. 8b.

Comparison of curves 1 and 2 in Fig. 8a shows that the reaction product of the copolymerization contains the PTBA-*g*-PS graft copolymers (peak at 23 min) together with unconverted PS-MA macromonomer (peak at

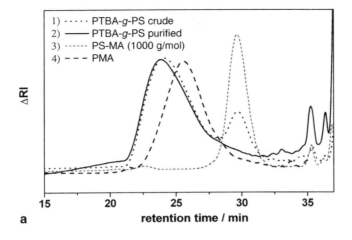

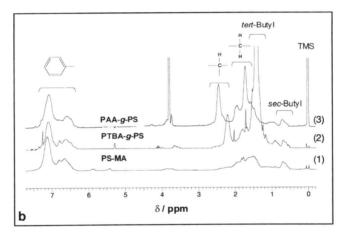

Fig. 8 (**a**) SEC curves of the PS macromonomer (PS-MA; *curve 1*), the crude (*curve 2*) and the purified (*curve 3*) poly(*tert*-butyl acrylate–*graft*-styrene) (PTBA-*g*-PS), and the esterified polymethylacrylate (PMA; *curve 4*) backbone. (**b**) ^{1}H NMR spectra of the macromonomer PSMA, the graft copolymer PTBA-*g*-PS and the final purified reaction product poly(acrylic acid-*graft*-styrene) (PAA-*g*-PS)

30 min). The unreacted PS-MA can be separated from the crude graft copolymer by silica gel column chromatography with dichloromethane and ethyl acetate [177a]. The SEC curve 3 of the purified graft copolymer shows the absence of the macromonomer peak and the unchanged molecular weight of the graft copolymer.

5.4
Characterization of Amphipolar Graft Copolyelectrolytes

The principle of analysis of the constitution of the graft copolymer is evident from the reaction scheme in Scheme 15. The exact characterization of the graft copolymers with regard to the ionic backbone molecular weight and graft density and copolymer composition is based on a combination of chemical and physical techniques [184]. The overall comonomer composition of PTBA-g-PS and of the copolyelectrolyte PAA-g-PS can easily be obtained by ^{1}H NMR spectroscopy, and the average graft length is given by the known molecular weight of the PS macromonomer. Information about the average poly(acrylic acid) (PAA) backbone length of the graft copolyelectrolyte is accessible by SEC analysis of the fully saponified copolymer. Saponification with KOH in dioxane at 100 °C is carried out first; under these conditions the methacryloyl ester group (carrying the PS graft) which is stable under the acidic conditions of the tert-butyl ester cleavage of the polyacrylate backbone (see above) is also cleaved [185]. The resulting reaction mixture contains both PAA and cleaved off hydroxy terminated polystyrene (PS-OH) grafts. The resulting PAA backbone polymer may be separated from the cleaved PS-OH grafts by extraction of the reaction mixture with aq. alkaline solution in order to avoid interfering elution signals in GPC analysis. After separation of the backbone from the cleaved grafts, the PAA is converted into poly(methyl acrylate) (PMA) by esterification with diazomethane (cf. Ref. [186]). This polymer analogous reaction is necessary in order to avoid erroneous results often obtained in polyelectrolyte molecular weight determination. Thus the molecular weight of the PAA backbone can be derived from the PMA backbone analogue by SEC using a PMA calibration; the SEC curve of this PMA is curve 4 of Fig. 8a.

Results from characterization of the constitution of the synthesized graft copolymers, compiled in Table 1, illustrate the versatility of the synthetic strategy outlined in the reaction scheme (Scheme 15) for the preparation and characterization of tailored graft copolyelectrolytes.

These studies have shown that well defined ionic–hydrophobic graft copolymers can be readily prepared by applying the macromonomer technique. The graft length is given by the macromonomer employed, and the backbone length is easily controlled by the use of benzyl mercaptan as chain-transfer reagent. The exact characterization of the graft copolymers with regard to the ionic backbone molecular weight and copolymer composition and graft density is only achieved by combination of chemical and physical analytical techniques. The graft copolymers obtained are suitable model systems for investigation of the potential of these copolymers as polymeric amphipolar stabilizers (cf. Ref. [187]).

Table 1 Graft copolymer constitution as determined by ^{1}H NMR and SEC experiments: Comonomer composition and molecular weight of the PS-grafts ($M_{n,graft}$), the PAA backbone length ($M_{n,BB}$), and the graft density (GD)

Copolymer	$M_{n,PS-MA}$ [a] of PS-MA	$\frac{[TBA]_0}{[PS-MA]_0}$ [b]	[aa] [c] (mol-%)	$M_{n,graft}$ (PD) [d] of PTBA-g-PS	$M_{n,BB}$ (PD) [e] of PAA	$M_{n,PAA}-g-M_{n,PS}$ [f] (GD) of PAA-g-PS
PAA-g-PS-1	1000	35:1	88	10.100 (1.54)	4300 (1.31)	4300–g–800 (0.8)
PAA-g-PS-2	1000	25:1	84	5400 (1.63)	3200 (1.23)	3200–g–900 (0.9)
PAA-g-PS-3	1000	17:1	76	7600 (1.58)	3400 (1.23)	3400–g–1500 (1.5)
PAA-g-PS-4	2000	75:1	84	8100 (1.52)	3300 (1.18)	3300–g–1000 (0.5)
PAA-g-PS-5	2000	50:1	70	6500(1.56)	2700 (1.16)	2700–g–1600 (0.8)
PAA-g-PS-6	2000	25:1	65	8300(1.61)	4300 (1.12)	4300–g–3200 (1.6)

[a] Number-average molecular weight $M_{n,PS-MA}$ of employed polystyrene macromonomer as determined by SEC analysis (PS calibration)

[b] Charged comonomer mole ratio

[c] Acrylic (acid) comonomer [aa] in graft copolymer as obtained from ^{1}H NMR analysis of poly(tert-butyl acrylate-graft-styrene) graft copolymer (PTBA-g-PS)

[d] Number-average molecular weight $M_{n,graft}$ and polydispersity (PD) of PTBA-g-PS as determined by SEC analysis (PS calibration)

[e] Number-average molecular weight $M_{n,BB}$ and polydispersity (PD) of poly(acrylic acid) (PAA) backbone polymer as derived from SEC analysis (polymethacrylate (PMA) calibration) of the PMA backbone analog (methyl esterified PAA of graft copolymer saponification)

[f] Molecular composition (number-average $M_{n,PAA}$ smaller g-$M_{n,PS}$) and PS graft density (GD) of PAA-g-PS graft copolymer as calculated from $M_{n,BB}$ and [aa]

6
Conclusions

Well-defined polyelectrolytes covering a wide variety of molecular architectures are available using precise polymerization techniques, for example anionic, controlled radical, and macromonomer polymerization of protected or reactive monomers, and the Suzuki reaction in a first step. The resulting well-characterized precursors were readily modified to the charged macromolecules. Variation of the chemical structure includes stiff-chain cationic and anionic polymers based on poly(p-phenylene), flexible cationic polyelectrolytes containing moieties of different hydrophobicity, polycarbobetaines with adjusted hindrance of the interaction of the ionic charges, star-shaped poly(acrylic acid) with broad variability of the number and the length of the arms, highly branched poly(acrylic acid) with regulated molecular parameters, ionic–hydrophobic graft and block copolymers with adjustable graft and backbone length in the former case and adjustable block length and block length ratio in the latter. These very different polyelectrolytes serve as models for investigation of the properties of ionically charged macromolecules in solution and their interaction with oppositely charged macroions, colloidal particles, or solid surfaces. Examples given in this chapter are the dependence of typical solution properties, for example viscosity, of the polyion–counterion interaction and of the electrophoretic mobility on the chemical structure and the molecular parameters of the polyelectrolytes involved. The amphiphilic block and graft copolymers are shown or proposed to be powerful electrosteric stabilizers for different colloids in aqueous systems.

Acknowledgement The authors are grateful to Professor Dr M. Ballauff, Dr J. Blaul, Dr G. Brodowski, Dr I. U. Rau, and Dr M. Wittemann, Universität (TH) Karlsruhe, C. Kozlowski, Dr U. Wendler, Dr T. Schimmel, Fraunhofer IAP (Golm), Dr Daniela Held, PSS GmbH, Mainz, Mingfu Zhang, Hans Lechner, Universität Bayreuth, Dr K. Dirnberger, Institut für Angewandte Makromolekulare Chemie, Universität Stuttgart, and Dr T. Schauer, Forschungsinstitut für Pigmente und Lacke e.V., Stuttgart, for their important contributions to the developments presented in this review article. For financial support, the authors wish to thank the Deutsche Forschungsgemeinschaft (DFG) and the Fonds der Chemischen Industrie (FCI).

References

1. Mandel M (1988) Polyelectrolytes In: Mark FH, Bikales NM, Overberger CG, Menges G (eds) Encyclopedia of polymer science and engineering, 2nd edn, vol 10. Wiley, New York, p 739
2. Schmitz KS (1993) Macroions in solution and colloid suspension. VCH, New York
3. Förster S, Schmidt M (1995) Adv Polym Sci 120:51
4. Auer HE, Alexandrowicz Z (1969) Biopolymers 8:1
5. Mandel M, Schouten J (1980) Macromolecules 13:11247
6. Nicolai T, Mandel M (1989) Macromolecules 22:438
7. Wang L, Bloomfield VA (1991) Macromolecules 24:5791

8. Kassapidou K, Jesse W, Kuil ME, Lapp A, Engelhaaf S, van der Maarel JRC (1997) Macromolecules 30:2671
9. Sato T, Norisuye T, Fujita H (1984) Macromolecules 17:2696
10. Gamini A, Mandel M (1994) Biopolymers 34:783
11. Milas M, Rinaudo M, Duplessix R, Borsali R, Lindner P (1995) Macromolecules 28:3119
12. Berth G, Dautzenberg H, Christensen BE, Harding SE, Rother G, Smidsrod O (1996) Macromolecules 29:3491
13. Schulz SF, Maier EE, Weber R (1989) J Chem Phys 90:7
14. Martin C, Kramer H, Johner C, Weyerich B, Biegel J, Deike R, Hagenbüchle M, Weber R (1995) Macromolecules 28:3175
15. Maier EE, Schulz SF, Weber R (1988) Macromolecules 21:1544
16. Maier EE, Krause R, Reggelmann M, Hagenbüchle M, Weber R (1992) Macromolecules 25:1125
17. Metzger Cotts P, Berry GC (1983) J Polym Sci Polym Phys Ed 21:1255
18. Lee CC, Chu SG, Berry GC (1983) J Polym Sci Polym Phys Ed 21:1573
19. Roitman DB, Wessling RA, McAlister J (1993) Macromolecules 26:5174
20. Roitman DB, McAlister J, McAdon M, Wessling RA (1994) J Polym Sci B Polym Phys 32:1157
21. a) Farmer BL, Chapman BR, Dudis DS, Adams WW (1983) Polymer 34:1588; b) Galda P (1994) Thesis, Karlsruhe; c) Rulkens R (1996) Thesis, Mainz; d) Vanhee S, Rulkens R, Lehmann U, Rosenauer C, Schulze M, Köhler W, Wegner G (1996) Macromolecules 29:5136
22. Rau IU, Rehahn M (1993) Macromol Chem 194:2225
23. Rau IU, Rehahn M (1993) Polymer 34:2889
24. Rau IU, Rehahn M (1994) Acta Polymerica 45:3
25. Brodowski G, Horvath A, Ballauff M, Rehahn M (1996) Macromolecules 29:6962
26. Wittemann M, Rehahn M (1998) J Chem Soc Chem Commun 623
27. Traser S, Wittmeyer P, Rehahn M (2002) e-Polymers #032,http://www.e-polymers.org
28. Balanda PB, Ramey MB, Reynolds JR (1999) Macromolecules 32:3970
29. Harrison BS, Ramey MB, Reynolds JR, Schanze KS (2000) J Am Chem Soc 122:8561
30. McQuade DT, Hegedus AH, Swager TM (2000) J Am Chem Soc 122:12389
31. Wallow TI, Novak BM (1991) J Am Chem Soc 113:7411
32. Wallow TI, Novak BM (1991) Polym Prepr Am Chem Soc Div Polym Chem 32:191
33. Wallow TI, Novak BM (1992) Polym Prepr Am Chem Soc Div Polym Chem 33:908
34. Wallow TI, Novak BM (2000) Polym Prepr Am Chem Soc Div Polym Chem 201:1009
35. Chield AD, Reynolds JR (1994) Macromolecules 27:1975
36. Kim S, Jackiw J, Robinson E, Schanze KS, Reynolds JR, Baur J, Rubner MF, Boils D (1998) Macromolecules 31:964
37. Rulkens R, Schulze M, Wegner G (1994) Macromol Rapid Commun 15:669
38. Vanhee S, Rulkens R, Lehmann U, Rosenauer C, Schulze M, Köhler W, Wegner G (1996) Macromolecules 29:5136
39. Rulkens R, Wegner G, Thurn-Albrecht T (1999) Langmuir 15:4022
40. Baum P, Meyer WH, Wegner G (2000) Polymer 41:965
41. Bockstaller M, Köhler W, Wegner G, Vlassopoulos D, Fytas G (2000) Macromolecules 33:3951
42. Bockstaller M, Köhler W, Wegner G, Vlassopoulos D, Fytas G (2001) Macromolecules 34:6359
43. Dautzenberg H, Jaeger W, Kötz J, Philipp B, Seidel Ch, Stscherbina D (1994) Polyelectrolytes: formation, characterization, application. Hanser, Munich
44. Cölfen H (2001) Macromol Rapid Commun 22:219
45. Bromberg L (2002) In: Tripathy SK, Kumar J, Nalioa HS (eds) Handbook of polyelectrolytes and their applications. American Scientific, p 1
46. Kathmann EE, White LA, McCormick CL (1997) Polymer 38:879
47. Bonte N, Laschewsky A (1996) Polymer 37:211

48. Niu A, Liaw DJ, Sang HC, Wu C (2000) Macromolecules 33:3492
49. Kathmann EE, White LA, McCormick CL (1997) Polymer 38:871
50. Kathmann EE, McCormick CL (1997) J Polym Sci A Polym Chem 35:243
51. Favresse P, Laschewsky A (2001) Polymer 42:2755
52. Ali SA, Rasheed A, Wazeer MIM (1999) Polymer 40:2439
53. Ali MM, Perzanowski HP, Ali SA (2000) Polymer 41:5591
54. Wielema TA, Engberts JBFN (1990) Eur Polym J 26:415
55. Ladenheim H, Morawetz H (1957) J Polym Sci 26:251
56. Barboiu V, Streba E, Luca C, Simionescu CI (1995) J Polym Sci A Polym Chem 33:389
57. Nagaya J, Uzawa H, Minoura N (1999) Macromol Rapid Commun 20:573
58. Favresse P, Laschewsky A, Emmermann C, Gros L, Linsner A (2001) Eur Polym J 37:877
59. Ali SA, Aal-e-Ali (2001) Polymer 42:7961
60. Cölfen H (2001) Macromol Rapid Commun 22:219
61. a) Selb J, Gallot Y (1980) Macromol Chem 181:809; b) Selb J, Gallot Y (1981) Macromol Chem 182:1513
62. Oh JM, Lee HJ, Shim HK, Choi SK (1994) Polym Bull 32:149
63. Lieske A, Jaeger W (1998) Macromol Chem Phys 199:255
64. Jaeger W, Wendler U, Lieske A, Bohrisch J (1999) Langmuir 15:4026
65. Lieske A, Jaeger W (1999) Tenside Surf Det 36:155
66. Zintchenko A, Dautzenberg H, Tauer K, Khrenov V (2002) Langmuir 18:1386
67. Ranger M, Jones M, Yessine M, Leroux J (2001) J Polymer Sci A Polymer Chem 39:3861
68. Wendler U, Bohrisch J, Jaeger W, Rother G, Dautzenberg H (1999) Macromol Rapid Commun 19:185
69. Bertin D, Boutevin B (1996) Polym Bull 37:337
70. Kazmeier PM, Daimon K, Georges MK, Hamer GH, Veregin RPN (1997) Macromolecules 30:2228
71. Lacroix-Desmazes P, Delair T, Pichot C, Boutevin B (2000) J Polym Sci A Polym Chem 38:3845
72. Bohrisch J, Wendler U, Jaeger W (1997) Macromol Rapid Commun 18:975
73. Fischer A, Brembilla A, Lochon P (1999) Macromolecules 32:6069
74. Jaeger W, Bohrisch J, Wendler U, Lieske A, Hahn M (1999) Novel synthetic polyelectrolytes. In: Noda I, Kokufuta E (eds) Proc Yamada Conference L Polyelectrolytes. Yamada Science Foundation, p 13
75. Jaeger W, Wendler U, Lieske A, Bohrisch J, Wandrey C (2000) Macromol Symp 161:87
76. Wandrey C, Hunkeler D, Wendler U, Jaeger W (2000) Macromolecules 33:7136
77. Malinova V, Schmidt B, Freitag R, Wandrey C (2002) Int Symp Polyelectrolytes "Polyelectrolytes 2002", Lund, 15–19 Jun, 2002
78. Bohrisch J, Grosche O, Wendler U, Jaeger W, Engelhardt H (2000) Macromol Chem Phys 201:447
79. Grosche O, Bohrisch J, Wendler U, Jaeger W, Engelhardt H (2000) J Chromatogr A 894:105
80. Bohrisch J, Schimmel T, Engelhard H, Jaeger W (2000) Macromolecules 35:4143
81. a) Schimmel T, Bohrisch J, Jaeger W (2002) unpublished work; b) Schimmel T (2002) Thesis. Technical University Berlin
82. Jaeger W, Paulke BR, Zimmernann A, Lieske A, Wendler U, Bohrisch J (1999) Polymer Prepr Am Chem Soc Div Polym Chem 40:980
83. Thünemann AF, Wendler U, Jaeger W (2000) Polym Int 49:782
84. Kozlowski C, Schimmel T, Bohrisch J, Jaeger W (2002) unpublished work
85. Izumrudov VA, Zelikin AN, Jaeger W, Bohrisch J, in preparation
86. Smits RG, Koper GJM, Mandel M (1993) J Phys Chem 97:5745
87. Borkovec M, Koper GJM (1998) Prog Colloid Polym Sci 109:142
88. Muthukumar M (1993) Macromolecules 26:3904
89. Wolterink JK, Leermakers FAM, Fleer GJ, Koopal LK, Zhulina EB, Borisov OV (1999) Macromolecules 32:2365

90. Borkovec M, Koper GJM (1997) Macromolecules 30:2151
91. Borisov OV, Birshtein TM, Zhulina EB (1992) Prog Colloid Polym Sci 90:177
92. Borisov OV, Zhulina EB (1998) Eur Phys J B 4:205
93. Misra S, Mattice WL, Napper DH (1994) Macromolecules 27:7090
94. Matyjaszewski K (2000) (ed) Controlled/living radical polymerization. Progress in ATRP, NMP, RAFT. ACS Symp Ser, vol 768
95. Qiu J, Charleux B, Matyjaszewski K (2001) Prog Polym Sci 26:2083
96. Matyjaszewski K, Xia J (2001) Chem Rev 101:2921
97. Kamigaito M, Ando T, Sawamoto M (2001) Chem Rev 101:3689
98. Hawker CJ, Bosman AW, Harth E (2001) Chem Rev 101:3661
99. Rizzardo E, Chiefari J, Mayadunne R, Moad G, Thang S (2001) Macromol Symp 174:209
100. Zioga A, Sioula S, Hadjichristidis N (2000) Macromol Symp 157:239
101. Pitsikalis M, Sioula S, Pispas S, Hadjichristidis N, Cook DC, Li J, Mays JW (1999) J Polym Sci A Polym Chem 37:4337
102. Hadjichristidis N, Pispas S, Pitsikalis M, Iatrou H, Vlahos C (1999) Adv Polym Sci 142:71
103. Voulgaris D, Tsitsilianis C, Grayer V, Esselink FJ, Hadziioannou G (1999) Polymer 40:5879
104. Hückstädt H, Göpfert A, Abetz V (2000) Macromol Chem Phys 201:296
105. Heinrich M, Rawiso M, Zilliox JG, Lesieur P, Simon JP (2001) Eur Phys J E 4:131
106. Smisek DL, Hoagland DA (1990) Science 248:1221
107. Tsitsilianis C, Lutz P, Graff S, Lamps JP, Rempp P (1991) Macromolecules 24:5897
108. Mengel C, Meyer WH, Wegner G (2001) Macromol Chem Phys 202:1138
109. Held D, Müller AHE (2000) Macromol Symp 157:225
110. Held D (2000) Dissertation, Johannes-Gutenberg-Universität Mainz
111. Heise A, Hedrick JL, Frank CW, Miller RD (1999) J Am Chem Soc 121:8647
112. Schnitter M, Engelking J, Heise A, Miller RD, Menzel H (2000) Macromol Chem Phys 201:1504
113. Zhang X, Xia J, Matyjaszewski K (2000) Macromolecules 33:2340
114. Fréchet JMJ, Henmi M, Gitsov I, Aoshima S, Leduc MR, Grubbs RB (1995) Science 269:1080
115. Baskaran D (2001) Macromol Chem Phys 202:1569
116. Simon PFW, Radke W, Müller AHE (1997) Macromol Rapid Commun 18:865
117. Hawker CJ, Fréchet JMJ, Grubbs RB, Dao J (1995) J Am Chem Soc 117:10763
118. Weimer MW, Fréchet JM, Gitsov I (1998) J Polym Sci A Polym Chem 36:955
119. Matyjaszewski K, Gaynor SG, Kulfan A, Podwika M (1997) Macromolecules 30:5192
120. Matyjaszewski K, Gaynor SG, Müller AHE (1997) Macromolecules 30:7034
121. Matyjaszewski K, Gaynor SG (1997) Macromolecules 30:7042
122. Sunder A, Hanselmann R, Frey H, Mülhaupt R (1999) Macromolecules 32:4240
123. Fréchet JMJ, Aoshima S (1996) WO Patent 9 614 346
124. Gaynor SG, Edelman S, Matyjaszewski K (1996) Macromolecules 29:1079
125. Simon PFW, Müller AHE (2001) Macromolecules 34:6206
126. Paulo C, Puskas JE (2001) Macromolecules 34:734
127. Mori H, Chan Seng D, Lechner H, Zhang M, Müller AHE (2003) Macromolecules 36:9270–9281
128. Sato Y, Hashidzume A, Morishima Y (2001) Macromolecules 34:6121
129. Cheng G, Simon PFW, Hartenstein M, Müller AHE (2000) Macromol Rapid Commun 21:846
130. Zhao B, Brittain WJ (2000) Prog Polym Sci 25:677
131. Kötz J, Kosmella S, Beitz T (2001) Prog Polym Sci 26:1199
132. Biesalski M, Rühe J (1999) Macromolecules 32:2309
133. Guo X, Ballauff M (2001) Phys Rev E 64:051406/1
134. de Robillard Q, Guo X, Ballauff M, Narayanan T (2000) Macromolecules 33:9109
135. Guo X, Ballauff M (2000) Langmuir 16:8719

136. Guo X, Weiss A, Ballauff M (1999) Macromolecules 32:6043
137. Pyun J, Matyjaszewski K (2001) Chem Mater 13:3436
138. Guerrini MM, Charleux B, Vairon JP (2000) Macromol Rapid Commun 21:669
139. Shah RR, Merreceyes D, Husemann M, Rees I, Abbott NL, Hawker CJ, Hedrick JL (2000) Macromolecules 33:597
140. Matyjaszewski K, Miller PJ, Shukla N, Immaraporn B, Gelman A, Luokala BB, Siclovan TM, Kickelbick G, Vallant T, Hoffmann H, Pakula T (1999) Macromolecules 32:8716
141. Bergbreiter DE, Tao G, Franchina JG, Sussman L (2001) Macromolecules 34:3018
142. Bergbreiter DE, Tao G (2000) J Polym Sci A Polym Chem 38:3944
143. Fujiki K, Sakamoto M, Sato T, Tsubokawa N (2000) J Macromol Sci Pure Appl Chem A37:357
144. Nakayama Y, Sudo M, Uchida K, Matsuda T (2002) Langmuir 18:2601
145. Hayashi S, Fujiki K, Tsubokawa N (2000) React Funct Polym 46:193
146. Mackay ME, Carmezini G, Sauer BB, Kampert W (2001) Langmuir 17:1708
147. Beyerlein D, Belge G, Eichhorn K-J, Gauglitz G, Grundke K, Voit B (2001) Macromol Symp 164:117
148. Voit B (2000) J Polym Sci A Polym Chem 38:2505
149. Mori H, Chan Seng D, Zhang M, Müller AHE (2002) Langmuir 18:3682
150. Mori H, Böker A, Krausch G, Müller AHE (2001) Macromolecules 34:6871
151. Zhou Y, Bruening ML, Bergbreiter DE, Crooks RM, Wells M (1996) J Am Chem Soc 118:3773
152. Peez RF, Dermody DL, Franchina JG, Jones SJ, Bruening ML, Bergbreiter DE, Crooks RM (1998) Langmuir 14:4232
153. Kim HJ, Moon JH, Park JW (2000) J Colloid Interface Sci 227:247
154. Reith D, Müller B, Müller-Plathe F, Wiegand S (2002) J Chem Phys 116:9100
155. Antonietti M, Weissenberger MC (1997) Macromol Rapid Commun 18:295
156. Clayton J (1998) Pigm Resin Technol 27:231
157. Tadros Th (1987) Solid/Liquid Dispersions. Academic Press, London
158. Napper DH (1983) Polymeric stabilization of colloidal dispersions. Academic Press, New York
159. Kaczmarski P, Tarng M, Glass JE, Buchacek RJ (1997) Prog Org Coat 30:15
160. Hoogeveen NG, Stuart MAC, Fleer GJ (1996), J Colloid Interface Sci 182:133
161. Schmitz J, Höfer R (1998) Farbe&Lack 104:22
162. Balazs AC, Siemasko CP (1991) J Phys Chem 95:3798
163. Van der Linden CC, Leermakers FAM, Fleer GJ (1996) Macromolecules 29:1000
164. Bijsterbosch HD, Stuart MAC, Fleer GJ (1998) Macromolecules 31:8981
165. Bijsterbosch HD, Stuart MAC, Fleer GJ (1999) J Colloid Interface Sci 210:37
166. Schaller Ch, Schauer T, Dirnberger K, Eisenbach CD (2001) Farbe&Lack 11:58
167. Milkovich RJ, Chiang MT (1974) US Patent 3 786 116
168. Schulz GO, Milkovich RJ (1982) J Appl Polym Sci 27:4773
169. Hadjichristidis N, Roovers JEL (1978) J Polym Sci Polym Phys Ed 16:851
170. Guanon Y, Grande D, Guerrero R (1998) Polym Prep Am Chem Soc Div Polym Chem 40:99
171. Roovers JEL (1976) Polymer 17:1107
172. Ma Y, Cao T, Webber TA (1998) Macromolecules 31:1773
173. Ishizu K, Mimmatsu S, Fukotoni T (1991) J Appl Polym Sci 43:2107
174. Milkovich R, Chiang MT (1974) US Patent 3 842 050 and subsequent patents
175. Schulz GO, Milkovich RJ (1984) J Polym Sci Polym Chem Ed 22:1633
176. O'Brien L, Gornick F (1955) J Am Chem Soc 77:4757
177. a) Schaller Ch (2002) Dissertation, University of Stuttgart; b) Schaller Ch (1998) Diploma Thesis, Stuttgart
178. Förster S, Abetz V, Müller AHE, Adv Polym Sci (this volume)
179. Tobolsky AV, Baysal B (1953) J Am Chem Soc 75:1757
180. Schaller Ch, Schauer T, Dirnberger K, Eisenbach CD (1999) Prog Org Coat 35:63
181. Masson P, Franta E, Rempp PF (1982) Macromol Rapid Commun 3:499

182. Quirk RP, Mathers RT, Wesdemiotis C, Arnould MA (2002) Macromolecules 35:2912
183. Schaller Ch, Schauer T, Dirnberger K, Eisenbach CD (2001) Eur Phys J 6:365
184. Haug PK, Eisenbach CD, J Polym Sci A, in preparation
185. Esselborn E, Fock J, Knebelkamp A (1996) Macromol Symp 102:91
186. Pizey JS (1974) Synthetic reaction, vol II. Halstedt, New York, chap 4
187. Reutter E, Silber S, Psiorz C (1999) Prog Org Coat 37:161

Received October 2002

Adv Polym Sci (2004) 165:43–78
DOI 10.1007/b11267

Poly(Vinylformamide-*co*-Vinylamine)/ Inorganic Oxide Hybrid Materials

Stefan Spange[1] · Torsten Meyer[1, 3] · Ina Voigt[1, 2] · Michael Eschner[2] · Katrin Estel[2, 3] · Dieter Pleul[2] · Frank Simon[2]

[1] Department of Polymer Chemistry, Chemnitz University of Technology, Strasse der Nationen 62, 09111 Chemnitz, Germany
  E-mail: *stefan.spange@chemie.tu-chemnitz.de*, E-mail: *t.meyer@leunapolymer.de*
[2] Institute of Polymer Research, Hohe Strasse 6, 01069 Dresden, Germany
  E-mail: *Ina-voigt@infineon.com*
  Present address:
[3] Semiconductor and Microsystems Technology Laboratory,
  Dresden University of Technology, 01062 Dresden, Germany

Abstract Novel ways to synthesize poly(vinylamine)/silica hybrid materials rich in free amino moieties using vinylformamide (VFA) as the key monomer are reported. Such materials are accessible from poly(vinylamine) which is obtained from radically produced poly(vinylformamide) (PVFA) which was either immobilized onto silica surfaces from solution and converted into poly(vinylamine-*co*-vinylformamide) polymers (PVFA-*co*-PVAm) by acidic hydrolysis, or by acidic hydrolysis in solution with subsequent adsorption on silica. We adsorbed PVFA-*co*-PVAm samples of different molecular masses and co-polymer compositions from aqueous solutions onto inorganic surfaces such as silica or titania simply by pH control. Direct surface functionalization using the VFA monomers was possible by their radical graft co-polymerization with bifunctional monomers, for example 1,3-divinylimidazolid-2-one (BVU) or on silica particles which were pre-functionalized with vinyltriethoxysilane (VTS). The influence of the amino content and molar mass (4000, 40,000, and 400,000 g mol^{-1}) of PVFA-*co*-PVAm on the degree of surface coverage, charging, and polarity was studied using X-ray photoelectron spectroscopy (XPS), potentiometric titrations, and electrokinetic measurements, and solvatochromic probes. It was shown that the amino content of the co-polymer has a significant influence on the amount of polymer adsorbed, the layer thickness, and surface polarity. Post-functionalization reactions with isocyanates or fullerene were used as a suitable method for enhancing the stability of the polyelectrolyte layer on the inorganic surface. They also open the way to producing multi-functional hybrid materials. The adsorption and post-functionalization of PVFA-*co*-PVAm samples onto gold-coated silicon wafer surfaces was used to build up laterally patterned surfaces for sensors and biological applications.

Keywords Hybrid material · Silica · Polyelectrolyte adsorption · Poly(vinylamine-*co*-vinylformamide)

Abbreviations and Symbols

ABAC	2,2'-Azobis-(2-amidinopropene) dihydrochloride
AFM	Atomic force microscopy
βb	Surface basicity parameter
BICDPM	4,4'-(Bisisocyanate)diphenyl methane
BVU	1,3-Divinylimidazolid-2-one
CP	Cross polarization
EPR	Electron paramagnetic resonance
IEP	Isoelectric point
HBD	Hydrogen bond donating
MAS	Magic angle spinning
n	Number of correlation points
PBVH	Poly(1,3-divinylimidazolid-2-one)
PVFA	Poly(vinylformamide)
PVFA-co-PBVH	Poly(vinylformamide-*co*-1,3-divinylimidazolid-2-one)
PVFA-co-PVAm	Poly(vinylformamide-*co*-vinylamine)
PVAm-co-PBVH	Poly(vinylamine-*co*-1,3-divinylimidazolid-2-one)
PVFA-co-PVAm-x	Poly(vinylformamide-*co*-vinylamine) which contains x w/w-% of the PVAm fraction (randomly distributed)
PVAm	Poly(vinylamine)
r	Correlation coefficient
σ	Standard deviation
SAM	Self-assembled monolayer
SANS	Small angle neutron scattering
VFA	Vinylformamide
VTS	Vinyltriethoxysilane

VTS-silica	Vinyltriethoxysilane grafted onto silica
ξ	Correlation length
XPS	X-ray photoelectron spectroscopy

In memory of Professor Dr. Hans-Jörg Jacobasch and to honour his work in surface sciences

1
Introduction and Motivation

The development of chemistry and materials sciences is divided into inorganic and organic chemistry. In different applications, for example metal coating, fibre-enhanced and filled polymers, adhesion, biomaterials, sensors and actor systems, the combination of inorganic and organic substances opened a successful means of obtaining multifunctional materials with improved properties for certain applications. The diversity of the possible combinations of substances and the demand for materials having novel exceptional properties has resulted in an increase in so-called inorganic-organic hybrid materials [1–4].

In this review we will discuss novel materials which preferably consist of silica and particular polyelectrolytes. Both components play important roles in nature and technical applications [5–8]. The advantage of silica and of numerous other metal oxides are due to their multifarious shape in which they can be found in natural sources or the well-defined shapes produced by means of different preparation techniques, e.g. spherical or angular grains with more or less random sizes, hollow spheres, nanoscale, mesoscopic or rather macroscopic particles with or without pores, transparent films and flat, more or less smooth, surfaces. Metal oxides are characterized by a defined shape (e.g. against heat or external pressure) and rather inert chemical stability. However, the latter can be also considered as disadvantageous for many technical applications. To perform the demand of mechanically stable but adequately reactive materials it seems satisfying to combine the advantageous properties of metal oxides with the high synthetic potential of polymers.

During recent decades a variety of synthetic concepts have been published using metal oxide surface groups (usually MeOH groups) as reactive anchors for low-molecular-weight organic components [9–12] and polymers [13–19]. It has been shown that under certain conditions metal oxide surfaces exhibit a reactivity which make them able to react directly with dissolved polymer coils [13–15] or dendrimers [16]. Another technique used to produce so-called oxide/polymer hybrids is the grafting of polymers or the performing of polymerization reactions on silanized metal oxide [17–26]. The surface-mediated polymerization of cationically polymerizable vinyl monomers, e.g. vinyl ethers, cyclopentadiene, furfuryl alcohol, *p*-methoxystyrene, etc., on silica surfaces is characterized by the formation of covalent Si–O–C bonds between the inorganic particle and the organic polymer [27–30].

While the covalent bonding of polymers on oxide surfaces requires certain reaction conditions adapted for the particular reaction system, for example highly purified non-aqueous solvents or thermally activated surface groups, surface modification with polyelectrolytes can be carried out in aqueous environment [31–46]. Numerous parameters can be adjusted to control the conformation of the polyelectrolyte molecules, e.g. their charge density and the charge density of the metal oxide surface. In other words, the adsorption process and the surface properties of the final product can be influenced in many different ways. This diversity is a challenge for academic research to develop novel hybrid materials for technical applications.

In the future hybrid materials made from selected polyelectrolyte species and silica will be applied to mimic biologically active surfaces by manipulation of their biological activity and compatibility or they will be used as support for enzyme catalysis (Scheme 1).

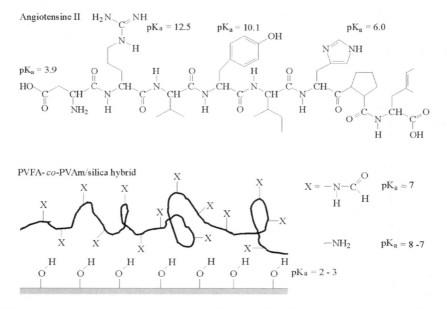

Scheme 1 Comparison between charges and Brønsted acid–base properties of the oligo-peptide angiotensin II and a PVFA-*co*-PVAm/silica interphase

Silica is a moderately strong solid acid, and in this sense it can also be considered as a polyelectrolyte, of course with a fixed geometry and non-conformational flexibility of the charge carriers. Controlled functionalization with a basic component, for example a polyelectrolyte, which is conformationally flexible, can be used to produce stable solid polymer hybrids with tailor-made surface properties for a lot of components [31–46].

However, many commercially available polyelectrolytes are chemically rather inert. Their saturated ionic charge carriers, like quaternary ammonium or carboxylate groups are not readily able to undergo further chemical derivatization or functionalization reactions in water. Primary amino moieties [47–49] are more suitable for dynamically adjustable pH control of the hybrid materials surface and various subsequent reactions can be carried out to fix the polyelectrolyte layer on the surface and to enhance the compatibility to proteins. The twenty-two amino acids of natural proteins build up versatile architectures with specific intramolecular interactions necessary for their functioning in biological processes. Only a few centers in the protein molecule are able to interact with the solid surface without a loss of the protein's molecular conformation and functioning. Hence, solids to be used for interaction with proteins or other biologically relevant liquids must be characterized by surface groups which are potential candidates for many different functionalization reactions. In this review, we describe the application of PVFA and PVAm co-polymers to modify metal oxide particles and flat wafer surfaces. Over silanes or other reactive polymers they have the advantage that they can be adsorbed from aqueous solutions, which are closest to all biological systems, avoid surface contamination with traces of artificial solvents, and help to guarantee environmental conservation [50–52].

2
Synthetic Concept and Experimental

2.1
Synthetic Concept

The monomer VAm is not available due to the well known tautomerization equilibrium with acetaldehyde imine. Therefore, polymer-analogous reactions must be employed to synthesize PVAm from suitable precursor polymers, e.g. poly(acrylamide) [53, 54], poly(vinyl acetamide) [55], poly(vinyl carbamate) [56], or poly(vinyl imides) [57]. Some of these polymer-analogous reactions require rather rigorous reaction conditions and side-products and degraded polymer chains disturb the properties of the final product.

A structurally well-defined and water-soluble PVAm is available by acid or base catalyzed hydrolysis of PVFA (Scheme 2) [58, 59] because formamide groups are readily hydrolyzed to amino groups. The degree of hydrolysis of PVFA can be used to produce adaptable copolymers, poly(vinyl formamide-*co*-vinyl amine) (PVFA-*co*-PVAm) with different formamide/amino group ratios. Hence, the degree of hydrolysis of the formamide groups is an excellent tool to control the desired charge density of the polyelectrolyte.

The molecular mass of PVFA-*co*-PVAm can be adjusted by the production of PVFA. High-molecular-mass polymers are yielded by radical polymerization of VFA [58, 59], while rather low-molecular-mass products (oligomers) can be obtained by specific cationic initiation reactions [60–62]. The differ-

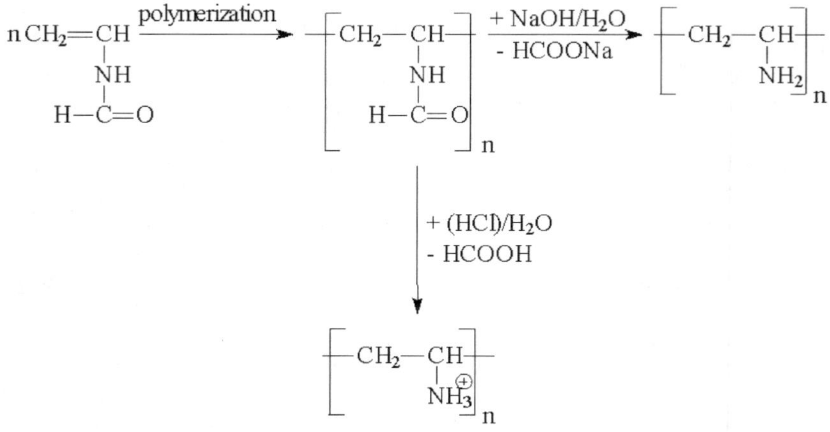

Scheme 2 Synthesis of PVFA-*co*-PVAm

ent synthetic routes allow preparation of PVFA with molecular masses over a wide range of M_n=800 g mol^{-1} to 400,000 g mol^{-1}.

Therefore, solely PVFA-*co*-PVAm obtained from VFA monomer has been considered in order to synthesize structurally well-defined hybrid materials by the procedures mentioned in Scheme 3. Immobilized or adsorbed PVFA-*co*-PVAm offers a high chemical potential to carry out subsequent reactions for several applications [63–66].

Because synthetic polymers and inorganic surfaces can interact with each other in many ways, various experimental procedures are available for synthesis and studies of polymers on surfaces of inorganic materials.

In this work, various synthetic concepts were developed to produce PVFA-*co*-PVAm/silica hybrid materials:

1. adsorption of PVFA-*co*-PVAm on inorganic particles;
2. subsequent functionalization of the adsorbed PVFA-*co*-PVAm;
3. cross-linking co-polymerization of VFA as suitable precursor monomer with bifunctional monomers on silica to produce PVAm networks (see below); and
4. grafting co-polymerization of VFA with immobilized vinylsilane (Scheme 3).

Sol–gel processes to synthesize polyelectrolyte/silica hybrid materials established in the literature [67, 68] were not pursued, because preliminary experiments showed that PVAm is unsuitable for this purpose [69].

Control of the ionic properties of the polyelectrolyte, e.g. charge density and acid–base strength, is possible by simple proton adsorption because the charge carriers, commonly –NH$_3$$^+X^-$ groups, are directly located along the polymer backbone. For numerous inorganic oxides, for example the silica or titania used, the adsorption of the PVAm can be carried out under condi-

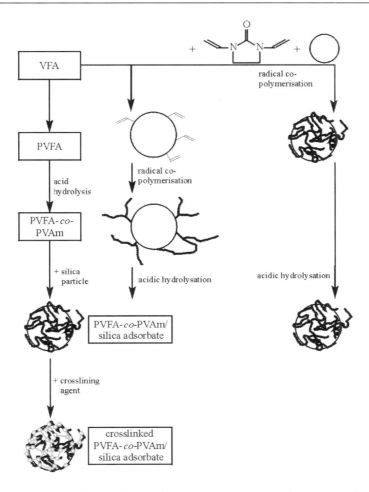

Scheme 3 Synthetic path to produce stable PVFA-*co*-PVAm/silica hybrid particles

tions where the polyelectrolyte molecules have a maximum of positive charges while the oxide surface is fully negatively charged as a result of the complete dissociation of the functional groups. A second favorable reason to use PVAm instead of polyelectrolytes bearing quaternized ammonium ions is that a large number of primary amino groups remain reactive after the surface modification. Hence, PVAm can be used for many consecutive chemical derivatization reactions [58, 59, 63, 64, 70].

Surface functionality of the hybrid material should be adjusted to different technological requirements. Cross-linking reactions, which enhance the stability of the coated material against mechanical and chemical attacks, allow build-up of polymer networks on the oxide material surface [63]. Finally, the reactive polyelectrolyte layer offers the opportunity for incorporation of nano-particles, for example noble metal clusters, metal oxide particles, or fullerenes [64, 66, 71]. The controlled incorporation of these substances makes the hybrid material suitable for build-up of sensor systems [66]. For this propose the essential build-up of a laterally well-defined polyelectrolyte structure seems possible by laterally controlled adsorption of PVFA-*co*-PVAm on silicon wafers covered with gold dots and pre-adsorbed mercapto-terminated carbonic acids [66].

2.2
Experimental

Spherical silica particles, for example Kieselgel (E. Merck, Darmstadt, Germany) and Aerosil (Degussa, Frankfurt am Main, Germany) are commercially available in a wide range of particle diameters, size distributions, and porosities. Microporous silica particles have pore distributions in the range between 40 and 60 Å. This means the majority of pores is inaccessible to the polyelectrolyte molecules but is accessible to small ions.

Precursor polymers (PVFA) of different molecular weight are also available commercially, e.g. from BASF AG (Ludwigshafen, Germany). Precursor oligomers have been synthesized by cationic polymerization of VFA with iodine as initiator [61]. PVFA-*co*-PVAm polyelectrolytes containing different amounts of amino groups (0 to 95% of the formamide groups of the PVFA used) were obtained by partial acid hydrolysis of the corresponding precursors with hydrochloric acid [60, 69]. The formation of block copolymers during the hydrolysis procedure can be excluded.

Usually, PVFA-*co*-PVAm samples were dissolved in aqueous KCl solutions of different ionic strength. Then, 0.5 g silica was suspended in 100 mL of the PVFA-*co*-PVAm solution. The adsorption equilibrium was reached after 24 h. The modified particles were filtered off, washed carefully with distilled water, and dried at room temperature.

The high amount of reactive amino groups on the surfaces of the hybrid materials allow post-functionalization reactions to be performed. Agents bearing only one group reactive towards amines, like epoxides, isocyanates, carbonic acids and their anhydrides and a second non-reactive group can be successfully used for an additional controlled surface functionalization. Agents having two or more groups of equal reactivity to amines offer the opportunity to cross-link the PVFA-*co*-PVAm chains to a stable network which wraps the silica particle.

Generally, neither the functionalization reaction nor the cross-linking reactions can be carried out in aqueous suspensions, because the coupling agents mentioned above react with water or are insoluble in it. Hence, water has to be completely exchanged by a liquid of low, e.g. toluene, benzene, or chlorinated hydrocarbons, or moderate donicity, e.g. acetone. These solvents allow functionalization or cross-linking under mild conditions [63, 64, 73, 74].

The situation is similar when conducting or semiconducting nano-particles are incorporated in the adsorbed polyelectrolyte layer. These particles can be noble metals to confer the hybrid material with catalytic properties [66, 71], metal sulfides [66], which can be involved in redox processes, or fullerenes [64], to control the electron pair donor–acceptor behavior to a suitable and appropriate substrate material, like titanium dioxide.

2.3
Characterization Methods

Study of the modification of solid surfaces requires, preferably, surface sensitive methods. Spectroscopic techniques, for example X-ray photoelectron spectroscopy (XPS) and FTIR spectroscopy are excellent tools for gathering information on the chemical surface composition and the kind and number of functional surface groups. The fact that the carbon and nitrogen containing organic phase is only introduced during the adsorption procedure and locally fixed on the outside of the particles allows the use of established methods for polymer and solid-state characterization, particularly NMR and solid-state NMR spectroscopy (e.g. ^{13}C CP MAS NMR).

Investigation of charges, charge densities and their changes during reactions, and characterization of surface polarity and reactivity (acid–base properties) requires methods describing the state of equilibrium between solid particles and surrounding aqueous media, for example potentiometric titration and electrokinetic measurements. Microelectrophoresis experiments were carried out with bare metal oxide or hybrid particles suspended in aqueous KCl solutions ($10^{-2} \geq c_{KCl} \geq 10^{-5}$) of different ionic strength and pH ($3 \geq pH \geq 10$; suspension: 0.01 g mL^{-1} molL^{-1} of particles). Zeta-potential values were calculated from the measured particle mobility. Potentiometric titrations to determine the surface charge densities of the solids were also carried out in diluted solutions of different KCl concentration and different pH.

3
Results and Discussion

3.1
Adsorption of PVFA-*co*-PVAm on Silica

The adsorption of PVFA-*co*-PVAm on silica was studied employing PVFA-*co*-PVAm samples of different degrees of hydrolysis and of different molecular weight (Table 1).

Silica particles suspended in water or aqueous solutions may be regarded as polyelectrolytes. The surface charge of the silica particles is either the result of dissociation processes of Brønsted acidic silanol groups (Si–OH) forming negatively charged silanolate ions (Si–O$^-$) or hydronium adsorption yielding Si–OH$_2^+$ species. Depending on pH, PVFA-*co*-PVAm molecules dissolved in aqueous electrolyte solution also bear charges on their chain. Hence, the Coulombic force between charges on the solid surfaces and on PVFA-*co*-PVAm chains can be considered an important component of the driving force for adsorption.

Table 1 Thickness of PVFA-*co*-PVAm layers adsorbed on to silicon wafers, determined by AFM, and IEP, determined by pH-dependent electrokinetic measurements in aqueous KCl solution (10^{-3} mol L^{-1}) using PVFA-*co*-PVAm/silica hybrid particles

Sample	Layer thickness (nm)	IEP[a]
Bare silica	0.00	03.50
PVFA-*co*-PVAm-0/silica	0.51	04.70
PVFA-*co*-PVAm-30/silica	0.53	09.25
PVFA-*co*-PVAm-48/silica	0.82	10.20
PVFA-*co*-PVAm-95/silica	1.13	10.75

[a] IEP=pH$|_{(\zeta\,=0)}$

Microelectrophoretic experiments on silica suspended in aqueous electrolyte solutions show that formation of negatively charged Si–O$^-$ ions starts at pH<3. At a pH≈9 all accessible Si–OH groups are dissociated [63]. The Brønsted acidity determined from electrokinetic data depends on the ionic strength of the KCl solutions. The pK_a values of the most acidic surface centers range between pK_a=4.4 and 5.6, which correspond to values of the adsorption molar free energies of the OH$^-$ ions (Φ_{OH}^-) of Φ_{OH}^-=−64.7 to −57.9 kJ mol^{-1} [73]. This indicates that in aqueous media the bare silica surface can be considered as a moderately strong Brønsted acid. With increasing pH the steady progress of silanol group dissociation can also be seen by an increase of the surface charge density σ_s (Fig. 1b).

In contrast to the results from electrokinetic experiments the curves of the potentiometric experiments do not show any approach to a plateau region observed in electrokinetic experiments at pH>9. The steady slope of the exponential function $\sigma_s=\sigma_s$ (pH) appears from taking the inner surface of the silica particle into account while electrokinetic experiments reflect only the charge distribution in the shear plane which is situated outside of particle surface.

The adsorption of the precursor polymer PVFA, which can be considered as the uncharged reference sample of PVFA-*co*-PVAm does not significantly influence the shape of the silica $\zeta=\zeta$ (pH) function (Fig. 1a). The shift of the IEP from pH=3 to pH=5 indicates that silanol groups are involved in the adsorption process (Table 1).

While proton transfer from silanol groups to the PVFA is not expected, it is assumed that these silanol groups are shielded by co-ordinatively bound PVFA segments or that they participate in hydrogen bonds with the polyelectrolyte molecules. The Brønsted acidity of the silica/PVFA surface is decreased to pK_a≈7. The conversion of 30% of the formamide groups into protonatable amino groups changes the zeta-potential–pH plot of the PVAm-30/silica hybrid surface to a typical shape indicating Brønsted basic surface properties (Fig. 1a). The protonated amino groups lead to positive zeta-potential values over a wide pH range up to IEP=8.8 (Table 1) and increasing the pK_a values to 8.5, corresponding to the pK_b of 5.5. With progressing hydrolysis of formamide groups the Brønsted base constant increases. With a

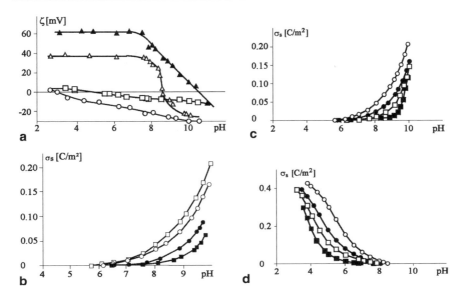

Fig. 1 (a) Influence of the degree of hydrolysis of the PVFA sample with M_n=400,000 g mol^{-1} on the zeta-potential as a function of pH, compared to bare silica measured in 10^{-3} mol L^{-1} KCl at 25 °C: bare silica (*open circles*), PVFA (*open squares*), PVFA-*co*-PVAm-30 (*open triangles*), PVFA-*co*-PVAm-95 (*filled triangles*); (b) Dependence of the surface charge, σ_s, of bare Kieselgel 60 on pH and ionic strength, measured in: pure water (*open squares*), 10^{-3} mol L^{-1} KCl (*open circles*), 10^{-2} mol L^{-1} KCl (*filled circles*), 10^{-1} mol L^{-1} KCl (*filled squares*); (c) Dependence of the surface charge, σ_s, of PVFA-*co*-PVAm-30/silica hybrid material on pH and ionic strength, measured in: pure water (*filled squares*), 10^{-3} mol L^{-1} KCl (*open squares*), 10^{-2} mol L^{-1} KCl (*filled circles*), 10^{-1} mol L^{-1} KCl (*open circles*); (d) Dependence of the surface charge, σ_s, of PVFA-*co*-PVAm-95/silica hybrid material on pH and ionic strength, measured in: pure water (*filled squares*), 10^{-3} mol L^{-1} KCl (*open squares*), 10^{-2} mol L^{-1} KCl (*filled circles*), 10^{-1} mol L^{-1} KCl (*open circles*)

pK_b of 4.9 the PVFA-*co*-PVAm-95/silica hybrid material is already a strong Brønsted base.

Despite the positive zeta-potential values of PVFA-*co*-PVAm-30/silica (Fig. 1a) and PVFA-*co*-PVAm-48/silica hybrid material over a pH range from 5.5 to 8.8 the corresponding surface charge densities calculated from results of potentiometric titrations are negative (Fig. 1c). Obviously, the number of positively charged amino groups does not compensate the negative net surface charge of the inner silica surface. The PVFA-*co*-PVAm chains with a rather low degree of hydrolysis are only suitable for adsorption on to a few sites of the silica surface; hence, a tail or a wide-loop conformation of the PVFA-*co*-PVAm chains should dominate. However, the tails and loops do not only cause the positive potential in the shear plane. The polyelectrolyte layer shifts the shear plane to greater distances from the rigid silica surface and decreases the value of the zeta-potential.

PVFA-*co*-PVAm/silica particles with PVFA-*co*-PVAm components bearing more than 50% of hydrolyzed formamide groups are able to fully overcompensate the negative net surface charge of the silica (Fig. 1d). The enormous number of interacting sites shifts the tail/loop amount of the adsorbed PVFA-*co*-PVAm chains to a higher amount of train conformation. As a result a more densely packed polyelectrolyte layer is built on the silica surface and shields it. The results show that train, loop, and tail conformations of adsorbed PVFA-*co*-PVAm chains can contribute to the particle surface structure, depending on the degree of charge localization along the PVFA-*co*-PVAm chain.

The adsorption of PVFA-*co*-PVAm on to silica was also studied by means of XPS. The XPS spectra of carefully purified silica, which is used as substrate material for PVFA-*co*-PVAm adsorption, does not contain substantial amounts of carbon (C 1 s) and nitrogen (N 1 s). After adsorption of PVFA-*co*-PVAm the amounts of carbon and nitrogen are increased significantly. The atomic ratio found experimentally to be $[N]:[C]_{exp}=0.473$ is close to the theoretically expected ratio of $[N]:[C]_{theo}=0.488$ for PVFA-*co*-PVAm with 95% of VAm groups. The high-resolution C 1 s spectrum of the PVFA-*co*-PVAm-95/silica hybrid material supports the assumption that the carbon peak appears only from the adsorbed polyelectrolyte molecules. The C 1 s spectrum was deconvoluted into three component peaks (Fig. 2a). Component peak *A* represents the hydrocarbon species C_xH_y. The second component peak *B* is shifted by 0.93 eV to higher binding energies. It reflects the presence of C –NH_2 and C –NH–C(H)=O bonds. A third component peak (*C*) has been found at 288.72 eV. This component peak appears from the presence of the formamide groups HN– C (H)=O of the former PVAF. The ratio of the component peak areas $[C]:[B]=0.04$ is in rather good agreement with the expected value at a degree of hydrolysis of 95% determined by ^{1}H NMR spectroscopy of PVFA-*co*-PVAm-95.

The influence of the molecular mass on the adsorbed PVFA-*co*-PVAm-95 fraction shown in Fig. 3, is detectable, in particular for pure aqueous solutions which do not contain KCl. Here, the amount of adsorbed PVFA-*co*-PVAm increases significantly with increasing molecular weight of PVFA-*co*-PVAm; when KCl is added to the aqueous solutions, however, the effect of molecular weight on adsorption behavior is not clearly detectable.

The decrease in the amount of polyelectrolyte adsorbed on the silica surface whilst increasing the salt concentration is in contrast to results obtained for adsorption of poly(diallyldimethylammonium chloride) on different silica samples [75, 76]. Probably the chloride ions shield the segment–segment interactions between charged groups inside the coil of an individual polyelectrolyte chain and the silica surface [77–80]. This explanation is also consistent with the assumption that electrostatic forces determine the adsorption mechanism of PVFA-*co*-PVAm chains on silica.

Furthermore the complete coverage of the silica surface was also evidenced by AFM scratch experiments of the polymer adsorbed onto flat silica surfaces (Fig. 4). These results match well with results reported by Akashi et

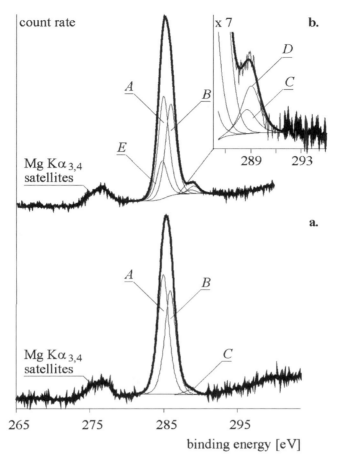

Fig. 2 C 1 s photoelectron spectra of PVFA-*co*-PVAm-95 (M_n=400,000 g mol^{-1}) adsorbed on to silica (**a**), and PVFA-*co*-PVAm-95 after crosslinking with BICDPM on silica (**b**)

al. [51], while angle-resolved XPS measurements show that poly(dial-lyldimethylammonium chloride) and other quaternary ammonium groups containing polyelectrolytes do not fully cover the silicon wafer [80].

The layer thickness of PVFA-*co*-PVAm-95 adsorbed on a silicon wafer was also estimated from angle-resolved XPS and ellipsometric measurements. From XPS measurements the layer thickness was determined to be 1.26 nm, and values obtained from ellipsometric measurements ranged between 1.0 and 1.5 nm [73]. These additional results correspond well with values determined from AFM scratch experiments (Table 1).

The E_T (30) polarity [81] of the PVAm/silica surface was determined with lipophilically substituted Reichardt's dye, 2,6-di(4-*tert*-butylphenyl)-4-[tris

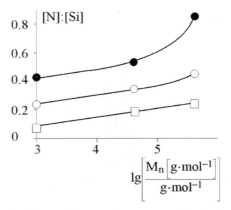

Fig. 3 Influence of molecular weight on the amount of adsorbed PVFA-*co*-PVAm-95 measured by XPS and indicated by [N]:[C] ratios. Adsorption experiments were carried out at pH=9 in: pure water (*filled circles*), 10^{-2} mol L^{-1} KCl (*open circles*), 5×10^{-2} mol L^{-1} KCl (*open squares*)

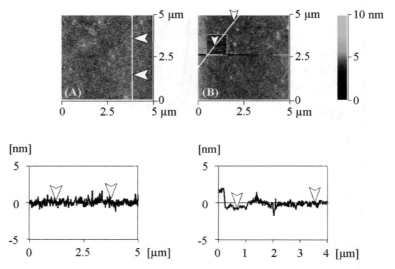

Fig. 4 AFM images of a bare silicon wafer (**A**), and a PVFA-*co*-PVAm-95 layer adsorbed on to a silicon wafer (**B**). In the area indicated in (**B**) the PVFA-*co*-PVAm layer has been removed by scratching

2,4,6-(4-*tert*-butylphenyl-*N*-pyridino)]phenolate [E_T (2)] (Scheme 4). According to its definition the original E_T (30) parameter is the molar transition energy corresponding to the longest wavelength of the UV/vis absorption of the standard dye 2,6-diphenyl-4-(2,4,6-triphenyl-*N*-pyridino)phenolate (**1**) (Eq. 1) [81, 82].

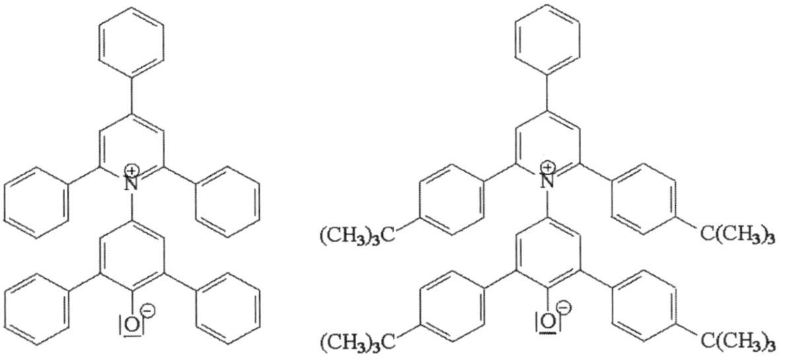

1   **2**

Scheme 4 Structure of the solvatochromic probe molecules **1** and **2**

$$E_T(30) = 2.891 \, \tilde{\nu}_{\max}(1) \cdot 10^{-3} \left[cm^{-1} \right] \tag{1}$$

When **2** is used instead of **1** then the E_T (30) value is given by Eq. (2) [82].

$$E_T(2) = 0.923 \, E_T(30) - 1.808 \tag{2}$$

Both dyes have been to be found suitable as surface polarity indicators for organically functionalized silica particles [83, 84]. In contrast to dye **1** the steric demand of the substituents in the 2- and 6-positions of the phenolate moiety prevent penetration of the dye **2** molecules through the organic layer towards the silica surface. Therefore, solutions of dye **2** in 1,2-dichloroethane or toluene seem very suitable as a probes to study the surface polarity of organically functionalized/silica hybrid particle [83].

As expected, the influence of the content of amino groups of the adsorbed PVFA-*co*-PVAm molecules on the polarity of the hybrid particles surface is fairly strong. Figure 5 shows the surface polarity parameters E_T (30) as function of the amino content of the polymer component of PVFA-*co*-PVAm/silica hybrid materials.

In comparison to the values obtained for toluene, the E_T (30) values calculated from measurements carried out in 1,2-dichloroethane are always about 3 kcal mol^{-1} larger. The difference is close to the expected value resulting from the polarity of the two solvents where E_T (30) of 1,2-dichloroethane is 41.3 kcal mol^{-1} and E_T (30) of toluene is 33.9 kcal mol^{-1} [81]. The decrease of E_T (30) values with increasing the content of amino groups in the adsorbed copolymer layer can be explained by a decrease of the HBD capacity and/or the dipolarity/dipolarizability of the former silica particle surface. Because the interaction between silanol groups and amino groups of PVFA-*co*-PVAm is really responsible for the decrease of the E_T (30) values, a relationship between the E_T (30) values and IEP of the PVFA-*co*-PVAm/silica hy-

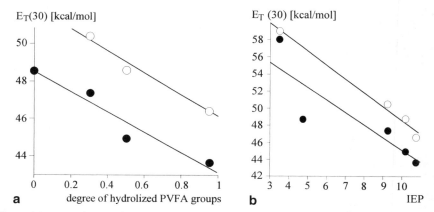

Fig. 5 (a) Dependence of E_T (30) values of several PVFA-*co*-PVAm/silica hybrid particles on the amino content in the PVFA-*co*-PVAm copolymer; E_T (30) values were determined from solvatochromic experiments using the adsorption of dye **2** from 1,2-dichloroethane (*open circles*), and toluene (*filled circles*); (b) E_T (30) values compared with the IEP values of several PVFA-*co*-PVAm/silica hybrid particles. E_T (30) values were determined from solvatochromic experiments using the adsorption of dye **2** from 1,2-dichloroethane (*open circles*), and toluene (*filled circles*). IEP data were determined by means of pH-dependent electrokinetic experiments in aqueous 10^{-3} mol L^{-1} KCl solution

brid particles was observed [85]. Also for organosilane-functionalized silica particles we found the linear relationship between IEP values determined from electrokinetic measurements in aqueous KCl solution and surface basicity parameter (β) determined by means of solvatochromic probes in 1,2-dichloroethane (Eq. 3) [85].

$$\beta = 0.2303 IEP - 1.09288 \tag{3}$$

with $n = 9$, $r = 0.989$, and $\sigma = 0.105$.

According to the symmetric acid–base theories, an increase of the basicity of the surface should cause a decrease of its acidity [86]. Hence, the increase of the β values should correspond to reduced E_T (30) values of the hybrid surface, because the silica surfaces HBD groups are involved in the adsorption process. Consequently, an inverse dependence of the E_T (30) on IEP values is expected. This is qualitatively shown in Fig. 5 for several PVFA-*co*-PVAm/silica hybrid particles.

3.2
Subsequent Reactions of PVFA-*co*-PVAm Adsorbed on Silica

To fix the adsorbed PVFA-*co*-PVAm layers irreversibly, cross-linking reactions with bifunctional or multifunctional reagents were carried out on the surface of solid PVFA-*co*-PVAm/hybrid particles. The cross-linking reactions were performed as two-step synthesis. In the first step PVFA-*co*-PVAm was

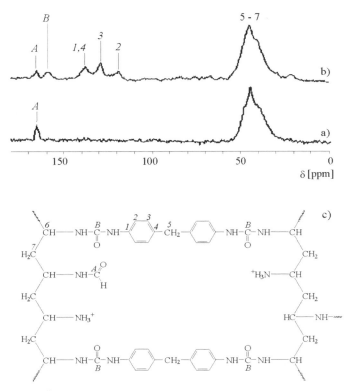

Fig. 6 Solid-state ^{13}C CP MAS NMR spectra of PVFA-*co*-PVAm-95/silica hybrid particles (M_n=400,000 g mol^{-1}) (**a**) before cross-linking, and (**b**) after cross-linking with BICDPM, (**c**) virtual structure of a BICDPM-functionalized PVFA-*co*-PVAm network cut out and assignment of the signals in the ^{13}C CPMAS NMR spectra

adsorbed from an aqueous solution, as described previously. In the second step the subsequent cross-linking reaction was carried out on to the separated particles using a suitable cross-linking agent dissolved in organic solvent (Scheme 3).

It is expected that these reactions, which occur between adsorbed PVAm sequences near the silica particle surface, transform the flexible polyelectrolyte layer into a more rigid one (Fig. 6). Agglomerate formation occurring between the PVFA-*co*-PVAm chains should be avoided.

3.2.1
Reaction with 4,4'-(Bisisocyanate)diphenylmethane (BICDPM)

We studied the cross-linking reaction of PVFA-*co*-PVAm adsorbed on to silica particles with (BICDPM) in acetone [63]. Isocyanates are very reactive compounds which react spontaneously not only with the primary amino groups but also with residual water, silanol surface groups, and formamide groups. Hence, the reaction of the (PVFA-*co*-PVAm)/silica hybrid particles with BICDPM may be accompanied by several side-reactions which disturb a selective functionalization process. Therefore, a relatively high content of the diisocyanate component is required in order to change the surface charge density.

The two isocyanate groups of BICDPM react with amino groups of the PVFA-*co*-PVAm chains as well as with silanol groups of the silica surface. Structure analysis of PVFA-*co*-PVAm/silica particles cross-linked with BICDPM was carried out by means of solid state ^{13}C CPMAS NMR spectroscopy (Fig. 6) and XPS (Fig. 2b).

Figure 6 shows the ^{13}C CPMAS spectrum of PVFA-*co*-PVAm-95/silica hybrid particles. Besides the broad signal of the main chain CH and CH$_2$ groups at about δ=46 ppm appearing from the adsorbed polyelectrolyte layer, an additional signal is observed at δ=165.2 ppm. This signal is caused by the carbonyl carbons of formamide groups remaining from the parent PVFA. The spectrum of this sample after reaction with BICDPM is shown in Fig. 6b. New signals appearing in the aromatic carbon region (δ=138.0, 128.9, and 119.2 ppm) and the shoulder at δ=41 ppm are due to the diphenyl methane moiety. The signal of the isocyanate group at about δ=125 ppm could not be detected; however, a new broadened carbonyl signal appears at δ=158.5 ppm. This chemical shift is expected for urethane and urea linkages [63]. This signal can also be observed for a reaction product of pure PVAm and BICDPM. Obviously, it is caused by the urea linkage between the PVAm backbone and the diphenylmethane unit, indicating the subsequent cross-linking reaction. Signals of a linkage between silanol groups and BICDPM are not detectable.

The NMR results confirm with the results of XPS measurements carried out on the cross-linked PVFA-*co*-PVAm-95/silica hybrid particles (Fig. 2b). The C 1 s spectrum is similar to the C 1 s spectrum of non-cross-linked PVFA-*co*-PVAm-95/silica hybrid material; however, the presence of the shoulder at about 288 eV required a C 1 s deconvolution into five component peaks. The component peaks *A*, *B*, and *C* represent the same structure elements discussed for the non-cross-linked PVFA-*co*-PVAm-95/silica hybrid. The new component peak *E* at BE=284.7 eV appears from carbon atoms of the phenyl ring of BICDPM. Component peak *D* at BE=289.03 eV shows the presence of urea groups formed by the cross-linking reaction of amino groups and isocyanate moieties. The area ratio [*D*]:[*E*]=0.1933 found experimentally is close to the expected ratio of 0.2. It shows that both isocyanate groups of BICDPM are able to react and cross-link the PVFA-*co*-PVAm-95 chains successfully. The introduction of two additionally amino groups and only one saturated hydrogen-bearing carbon atom per BICDPM molecule increases the ratio [*B*]:[*A*] slightly while [*C*]:[*A*] is nearly constant.

The post-cross-linking reaction of the amino groups with the isocyanate groups of BICDPM decreases the number of protonable amino groups in the PVFA-*co*-PVAm chains. As expected, with increasing amount of cross-link-

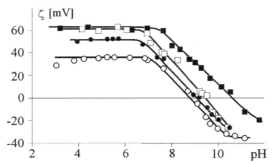

Fig. 7 Dependence of the zeta-potential of PVFA-*co*-PVAm-95/silica particles fuctionalized with different amounts of BICDPM on pH measured in aqueous 10^{-3} mol L^{-1} KCl at 25 °C: nonfuctionalized PVFA-*co*-PVAm-95/silica particles (*filled squares*), 0.01 w/w-% BICDPM (*open squares*), 0.05 w/w-% BICDPM (*filled circles*), 0.10 w/w-% BICDPM (*open circles*) (w/w-%=weight of BICDPM per gram PVFA-*co*-PVAm-95/silica·100%)

ing agent, corresponding to increased conversion of amino groups, the IEP of the hybrid particles surface shifts to lower pH values (Fig. 7).

Of course, the decrease in the number of positive charge carriers in the polyelectrolyte layer results in a decrease in the number of electrostatic interactions between the negatively charged silica surface centers and polyelectrolyte molecules. Advantageously, despite the limitation of electrostatic interactions of the polyelectrolyte layer and silica, the PVFA-*co*-PVAm cannot be removed because the cross-linking process forms a cage-like network covering the silica particle.

Despite the transformation of a certain number of amino groups into urea groups, the qualitative shape of the zeta-potential plots as a function of pH is not changed significantly. That indicates that PVFA-*co*-PVAm can be used as a precursor polymer for the gradual functionalization of silica particles and other Brønsted-acidic surfaces with amino functionality. Chemical reactions of PVFA-*co*-PVAm/silica particles with isocyanates or other electrophilic reagents offer a wide field of applications, because of the simple experimental procedure; this material combination seems promising for the construction of tailor-made polyelectrolyte networks on Brønsted-acidic surfaces or for other multicomponent systems.

3.2.2
Functionalization of PVFA-co-PVAm/Silica with Fullerenes (C60)

Fullerenes and their derivatives are of broad interest in various fields ranging from ferromagnetism [87] over their application as possible HIV inhibitors [88] to tumor-therapeutic active substance in biological systems [89]. Although C_{60} is insoluble in water, dissolution may be accomplished by using water-soluble polymers [90] or surfactant solutions containing amphiphilic block-copolymers [91], micelles or liposomes [92, 93]. The immobilization of

C_{60} on a polyelectrolyte layer is therefore also a new approach for studying interactions of C_{60} with biologically active compounds in water.

It is well known that C_{60} neither reacts with nor is adsorbed by bare silica, but it reacts with Si–H bonds on H-terminated silicon wafers forming covalently bonded monolayers [94]. The introduction of C_{60} dissolved in toluene should occur in the interfacial region between the PVFA-*co*-PVAm chains which are adsorbed on to silica and transform the reversibly adsorbed and flexible polyelectrolyte layer into an irreversibly adsorbed and more rigid structure. However, amino groups inside the adsorbed PVFA-*co*-PVAm coils are not available for surface functionalization with C_{60} (Scheme 5). Never-

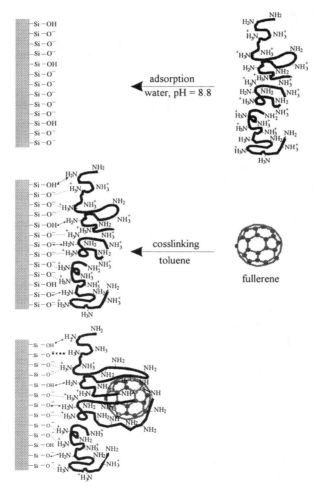

Scheme 5 Synthetic pathway towards stable cross-linked PVFA-*co*-PVAm/inorganic oxide hybrid particles

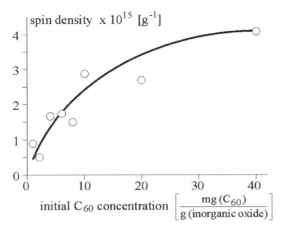

Fig. 8 Spin density of C_{60}/PVFA-*co*-PVAm/silica hybrid particles determined by means of EPR as a function of the initial C_{60} concentration used for cross-linking reactions. Details are given in Ref. [64]

theless, it will be shown that relatively small amounts of C_{60} influence the surface properties of the PVFA-*co*-PVAm/silica hybrid particles, such as their electrostatic properties and ability to undergo Coulombic interactions, by shielding surface charges.

The globular molecule C_{60} (diameter 1003 pm) is itself a nonpolar molecule and dispersion forces predominate in its interaction with the environment. For the subsequent reactions of PVFA-*co*-PVAm/silica hybrids with C_{60}, carefully dried samples were suspended in dry toluene. However, traces of water which are incorporated in the hydrophilic polyelectrolyte layer or adsorbed on bare silica cannot be excluded completely.

Fullerene C_{60} is an effective electron acceptor molecule which is able to form stable anion radicals by electron transfer from electron-donor molecules [95]. These radicals can be observed in EPR experiments. Of course, because of the small amount of C_{60} molecules incorporated in the polyelectrolyte layer only a small number of C_{60}^{-} anion radicals and very weak EPR absorption are detectable. Figure 8 shows that on increasing the initial C_{60} concentration used for the functionalization reaction the content of C_{60} on PVFA-*co*-PVAm/silica and PVFA-*co*-PVAm/titanium dioxide hybrid particle surfaces is at first increased and then approaches a limit of conversion.

According to Fig. 8, the radical concentration is about 2 to 5×10^{13} spins per gram of hybrid particles. It corresponds to an amount of about 10^{-8} mol of reacted C_{60} per gram hybrid material. Because of the small amount of C_{60} incorporated, it was expected that the post-cross-linking reaction of the amino groups of the PVFA-*co*-PVAm chains with C_{60} would not significantly decrease the amount of protonable amino groups. Surprisingly, electrokinetic measurements show that a small amount of C_{60} already strongly influ-

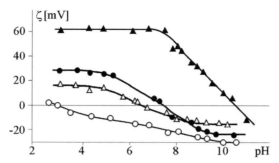

Fig. 9 pH dependence of the zeta-potential of bare silica, PVFA-*co*-PVAm/silica, and C$_{60}$/PVFA-*co*-PVAm/silica hybrid particles in aqueous KCl solution: bare silica in 10^{-3} mol L^{-1} KCl (*open circles*), PVFA-*co*-PVAm-95/silica hybrid particles in 10^{-3} mol L^{-1} KCl (*filled triangles*), C$_{60}$/PVFA-*co*-PVAm-95/silica hybrid particles in 10^{-3} mol L^{-1} KCl (*filled circles*), and C$_{60}$/PVFA-*co*-PVAm-95/silica hybrid particles in 10^{-2} mol L^{-1} KCl (*open triangles*)

ences the acid–base properties of the whole C$_{60}$-functionalized hybrid particle.

The strong shift of the IEP observed in the electrokinetic experiments carried out with C$_{60}$-modified PVFA-*co*-PVAm/silica hybrid particles (Fig. 9) indicates that approximately up to 30% of the former amino groups of the PVFA-*co*-PVAm chains must be involved in the interaction with incorporated C$_{60}$ molecules. This is a much higher degree of involved or converted amino groups than that observed for crosslinking reactions with large amounts of the polar diisocyanates. That result cannot be explained by disappearance of amino groups during the reaction with C$_{60}$. It is assumed that in presence of C$_{60}$ the weakly bonded and flexible PVFA-*co*-PVAm chains are able to undergo re-orientation processes. The force driving these processes is the rather strong and long-ranging dispersion force of C$_{60}$. The large polarizable C$_{60}$ molecules deform the PVFA-*co*-PVAm chains as suggested in Scheme 5.

AFM investigations of unmodified and C$_{60}$-functionalized PVFA-*co*-PVAm layers on a flat silica layer–deposited on a silicon wafer–are shown in Fig. 10. The fullerene functionalized PVFA-*co*-PVAm/silica layer causes holes with a diameter of 10 nm and depth of 2 nm. The former PVFA-*co*-PVAm layer has a thickness of about 1.2 nm. This shows the strong influence of reacted C$_{60}$ molecules on the conformation of the adsorbed polymer layer. These holes observed are also an indication that fullerene aggregates are encapsulated in the PVFA-*co*-PVAm layer on silica.

This interpretation also agrees with previously reported results of solvatochromism of C$_{60}$ in binary pyridine/water mixtures [96]. We assume that the amine groups of the PVAm encapsulate the fullerene molecules in a manner comparable with the behavior of pyridine molecules in water. The broad UV/Vis absorption spectra of the C$_{60}$-containing hybrids give an additional hint of aggregation of fullerene molecules on the PVFA-*co*-PVAm layer [64].

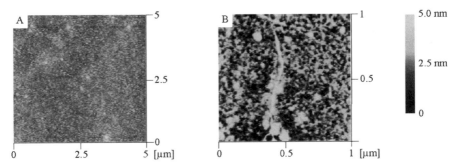

Fig. 10 AFM images of a PVFA-*co*-PVAm-95/silica (**a**) and a C_{60} functionalized PVFA-*co*-PVAm-95/silica (**b**) flat surface

The proposed deformation of the PVFA-*co*-PVAm layer and the incorporation of C_{60} do not influence significantly the shape of zeta-potential as function of pH (Fig. 5). This shows that the charge generation mechanism is not changed by the incorporated C_{60}. In other words, C_{60} does not influence the silica surface groups responsible for OH^- adsorption.

3.3
PVFA-*co*-PVAm Adsorbed on Gold-Functionalized Silica

Well defined multilayer films of organic compounds on solid surfaces for tailored architectures have been studied for more than 60 years. While the Langmuir–Blodgett technique is used to obtain molecular thin films of a high molecular order on surfaces, Decher [32] and Möhwald et al. [33, 34] extended chemisorption from solution to polyelectrolytes; this allows the construction of charged multilayer films. Such films from polyelectrolytes bear functional groups that can bind to other molecules, e.g. enzymes, antibodies, redox-active compounds, metal ions, etc. Metal ions, for instance, complexed by free amino groups will render the overall surface charge of the polyelectrolyte. This change in surface potential can be monitored by electrochemical methods allowing the monitoring of metal ions. Binding or embedding of redox-active organic compounds or metal clusters, e.g. FeS, provides the opportunity for the development of redox-active sensors which are related to biological redox-enzymes. Such systems are expected to be far more robust towards solvents and chemicals than enzymes. In other words, polyelectrolytes based on PVFA-*co*-PVAm seem decidedly suited to building up of sensor systems with laterally well defined polyelectrolyte layers. To transduce the information of the more or less specific reaction taking place in the functionalized or doted polyelectrolyte layer into electrical signals the polyelectrolyte should be placed in the immediate neighborhood of noble metals, e.g. gold, which can be act as electrodes. Those supramolecular architectures can be considered as triple hybrid materials which consist of an inorganic substrate, e.g. a thermally oxidized silicon wafer with or without a

TiO$_2$ coating, partially covered with gold dots and topped by a polyelectrolyte layer in which noble metal or FeS clusters are incorporated.

The adsorption of the polyelectrolytes was laterally controlled by an intermediate layer of ω-mercaptoalkyl carboxylic acids with different alkyl chain
lengths (Scheme 6).

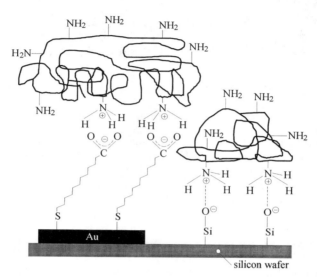

Scheme 6 Assumed structure of PVFA-*co*-PVAm tethered to surface bound 11-mercap-
toundecanoic acid

These ω-mercaptoalkyl carboxylic acids were selectively adsorbed from
ethanol solution (1 mmol L^{-1}) only on the gold dots which were laterally deposited in the first step on the silicon wafer. Imaging-XPS analysis revealed
the success of the laterally controlled adsorption of the ω-mercaptoalkyl carboxylic acids which can only be seen on areas coated by gold islands and
not on the silica bars in between (Fig. 11).

AFM offers the opportunity to investigate much smaller areas than with
XPS and ellipsometry. From those studies we conclude that 11-mercap-
toundecanoic acid does not form a dense layer on the gold surface [66]; instead this compound prefers to form clusters.

In aqueous solutions the carboxylate groups of the grafted mercapto components can act as anchors for PVFA-*co*-PVAm molecules. The adsorption
from differently concentrated PVFA-*co*-PVAm solutions (0.3 μmol L^{-1} to
3000 μmol L^{-1}) gave polyelectrolyte layer thicknesses between 0.9 and
2.7 nm (determined by ellipsometry and angle-resolved XPS). The values
found are in good agreement with the layer thicknesses on bare silicon wafer
surfaces [97]. The layer thicknesses of the pre-adsorbed mercapto compounds range from 0.5 to 0.9 nm. That means that at very low polyelectroyte
concentrations in the stock solution no increase in layer thickness was ob-

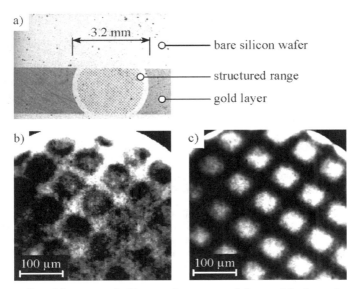

Fig. 11 Laterally gold structured silicon wafer to control the spatial adsorption of PVFA-*co*-PVAm: (a) photograph of the gold structured silicon wafer (genuine size of the gold islands 45 µm by 45 µm), (b) spatially resolved XPS image of the lateral gold layer (the Au 4f peak is depicted in *black* whilst the peaks of O 1 s and Si 2p are *pale*), (c) spatially resolved XPS image of 11-mercaptoundecanoic acid adsorbed on the laterally structured gold layer (the S 2p peak is depicted in *white* whilst Si 2p is *dark*)

served. At higher concentrations a pronounced increase in the layer thickness was measured. The exact nature of the obtained structures is still under investigation.

Binding of the polyelectrolyte molecules most likely occurs via salt formation between the carboxylic acid moiety and the free amino groups (Scheme 6). As confirmed by XPS analysis the polyelectrolyte molecules are not only adsorbed on gold islands but to a smaller extent on the silicon bars, also (Fig. 12a). But XPS-imaging revealed that the polyelectrolyte does not form a dense layer on the silica surface but small dispersed islands instead. Binding of the polyelectrolytes to the silica might occur via hydrogen-bond formation between the silanol groups on the silica surface and the free amino moieties of the polyelectrolytes. Surface hydrophobization of the bare silica areas with chloro-3,3,3-trifluoropropyldimethylsilane distinctly lowers the amount of polyelectrolyte adsorbed but does not fully prevent the polyelectrolyte from binding to the silicon bars (Fig. 12b).

For spatial reasons it is impossible that each amino group binds to a carboxylic acid moiety [97]. It should be also kept in mind that polymers do not usually form long chains but balls instead. Therefore not all of the amino groups in the polyelectrolyte are available for binding to the carboxyl moieties of the anchor groups. Consequently there must be free amino groups which are available for chemical modification or the incorporation

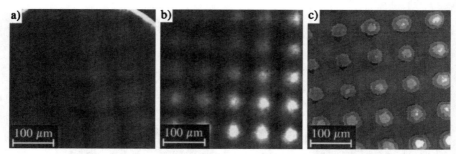

Fig. 12 (a) Spatially resolved XPS image of the lateral distribution of the PVFA-*co*-PVAm-95 layer on gold/11-mercaptoundecanoic acid-structured silicon wafer surfaces. The silicon wafer areas which were not coated by gold were not modified with hydrophobic silanes. The Si 2p peak is depicted in *black* whilst the peaks of N 1 s are *pale*; (b) Spatially resolved XPS image of the lateral distribution of the PVFA-*co*-PVAm-95 layer on gold/11-mercaptoundecanoic acid-structured silicon wafer surfaces. The silicon wafer areas which were not coated by gold were hydrophobized by means of chloro-3,3,3-trifluoropropyldimethylsilane. The Si 2p peak is depicted in *black* whilst the peaks of N 1 s are *pale*; (c) Spatially resolved XPS image of the lateral distribution of silver ions (the Ag 3d peak is *pale*) in the PVFA-*co*-PVAm-95 layer which was adsorbed on to a gold/11-mercaptoundecanoic acid-structured silicon wafer surface. The silicon wafer areas which were not coated by gold were hydrophobized by means of chloro-3,3,3-trifluoropropyldimethylsilane. The Si 2p peak is depicted in *black*

of metal-containing clusters or particles. The [N]:[S] ratios obtained from XPS measurements confirm our assumption on the presence of free amino groups. These free amino moieties are suitable anchor-groups for other molecules, e.g. redox-active compounds, enzymes, or antibodies, and bioconjugates, or deal as complexation agents for inorganic compounds. We successfully introduced $FeCl_2$ into the polyelectrolyte layer and were also able to convert the complexed iron to a mixture of FeS and FeS_2. Furthermore, we adsorbed silver ions from 1.0 mmol L^{-1} aqueous solution. The resulting complexes of polyelectrolyte and silver ions are analogous to the well known silver diamine complex [98]. XPS-imaging yielded images showing areas of high silver concentration which are identical with the positions of the gold islands (Fig. 12c). These experiments also confirm the presence of free amino groups in the surface-bound polyelectrolyte.

3.4
Grafting of PVFA-*co*-PVAm on Silica

By following the known synthetic route to cationically polymerized VFA [61] it should also have been feasible to co-polymerize this monomer; we showed, however, that co-polymerization of VFA in the presence of silica was impossible for mechanistic reasons [62]. Therefore we co-polymerized VFA radically using two different synthetic strategies. In the first we polymerized VFA with 1,3-divinylimidazolidin-2-one (BVU) as co-monomer in the presence of unmodified silica. BVU can be considered as an ideal co-monomer because the cross-linked copolymer network is stable towards the subsequent acidic hydrolysis which is inevitable if bisacrylamides are used as co-monomers. The second synthetic route uses vinyl silane-functionalized silica particles; the pre-grafted vinyl silane is used as co-monomer.

3.4.1
Radical Copolymerization of VFA with BVU in Presence of Silica Particles

BVU has been found to be a suitable cross-linking reagent (0.1 to 2.0 mol-%) for synthesis of PVFA/silica and PVFA-*co*-PVAm/silica hybrids by radical polymerization [72]. The synthetic approach is shown in Scheme 7.

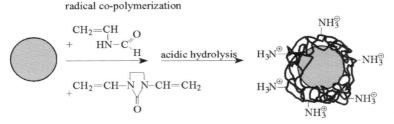

Scheme 7 Synthetic pathway to produce PVFA-*co*-PBVH and PVAm-*co*-PBVH networks on silica particle surfaces by radically initiated co-polymerization of VFA and BVH in the presence of silica

The formation of gel-like hybrid material significantly depends on the amount of the BVU used in the polymerization. Figure 13 shows the co-polymer content of PVFA-*co*-PBVU/silica hydrogels, expressed as the dependence of the total amount of carbon on the initial BVU content of the reaction mixture.

A minimum concentration of about 1.0 mol-% BVU in the reaction mixture was found to be necessary in order to cross-link the silica particles and form the desired hydrogel. Smaller quantities of BVU are not sufficient for effective cross-linking between the silica particles. With increasing initial amounts of BVU, saturation in the carbon content of the PVFA-*co*-PBVU/silica hydrogel of 25 w/w-% is observed. The largest amount available corresponds to 50 w/w-% VFA in the whole gel, indicating the complete formation of the PVFA-*co*-PBVU/silica hydrogel from the reaction mixture containing the silica particles and VFA in a 1:1 mass ratio.

Figure 14 shows a typical solid-state ^{13}C CP MAS NMR spectrum of a PVFA-*co*-PBVU/silica hydrogel sample selected from point B in Fig. 13. The two main signals at $\delta=42$ ppm and $\delta=164$ ppm correspond to the –CH$_2$– and –CHO structural units of PVFA. The small shoulder at $\delta\approx50$ ppm can be assigned to –CH– units. The signal observed at $\delta=100$ ppm is a spinning side

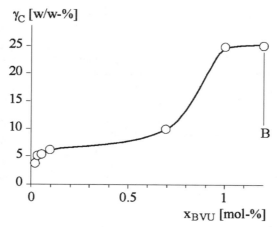

Fig. 13 Dependence of the total carbon content (γ_C) of the PVFA-*co*-PBVU/silica hydrogels on the initial BVU content of the reaction mixture (x_{BVU}). The carbon content was determined by combustion elemental micro-analysis

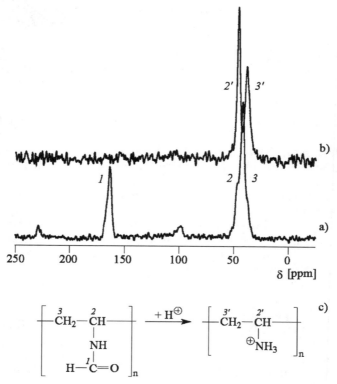

Fig. 14 Solid-state ^{13}C CP MAS NMR spectra of PVFA-*co*-PBVU/silica hydrogel before (**a**) and after 6 h hydrolysis with 18% HCl at 60 °C (**b**) and the reaction scheme for acidic hydrolysis with signal assignment (**c**)

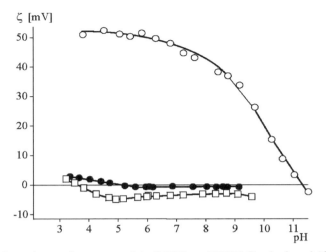

Fig. 15 pH-dependence of zeta-potential of PVFA-*co*-PBVU/silica hydrogel (*filled circles*) and its hydrolyzed species PVAm-*co*-PBVU/silica hydrogel (*open circles*), compared to bare silica (*open squares*), measured in 10^{-3} mol L^{-1} KCl at 25 °C

band. However, no signals derived from the small amount of BVU were detected in the ^{13}C CP MAS NMR spectrum.

To produce PVAm-*co*-PBVU/silica hydrogels the PVFA-*co*-PBVU/silica samples can be hydrolyzed either by acid or base [100]. To prevent dissolution of the silica and the decomposition of the hybrid under basic conditions acidic hydrolysis was applied. The success of the hydrolysis reaction over 6 h in 18% HCl at 60 °C can be seen in the solid state ^{13}C CP MAS NMR spectrum (Fig. 14b). The spectrum shows complete conversion of the formaldehyde groups into primary amino moieties. The broad signals of the –CH– and –CH$_2$– groups at δ=43 ppm in the solid-state ^{13}C CP MAS NMR spectrum of the former PVFA-*co*-PBVU/silica hydrogel is split into two signals appearing at δ=38 ppm and δ=46 ppm and derived from the –CH– and –CH$_2$– groups of PVAm. These NMR results are in complete agreement with literature data from soluble PVAm [58]. The carbon content of the hydrolyzed samples is approximately 12 w/w-%. According to results from PVAm-functionalized silica particles [63], a significant change in the surface properties of the hybrid is expected with release of amino groups. Figure 15 shows the pH-dependence of the zeta-potential of PVFA-*co*-PBVU/silica hydrogel and PVAm-*co*-PBVU/silica hydrogel samples compared to bare silica.

After cross-linking of the VFA and BVU on the silica particle surface the IEP shifts from pH≈3 (bare silica) to pH≈5 (PVFA-*co*-PBVU/silica). The shape of the function ζ=ζ (pH) remains unchanged. It can be assumed that the silanol groups on the silica surface are weakly shielded by the adsorbed sections of the cross-linked polymer because covalent attachment of the polymer does not take place [63, 73]. Strong effects can be seen after conversion of the formamide groups into amino groups. The IEP is shifted into the

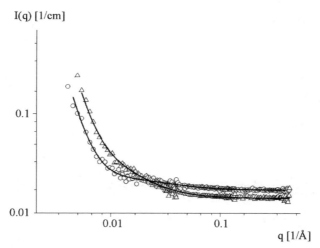

Fig. 16 SANS curves for PVFA-*co*-PBVU/silica (*open circles*) and PVAm-*co*-PBVU/silica hydrogels (*open triangles*) measured at T=298 K. The *solid lines* represent the fit according to Eq. (2)

basic region at pH>11. At lower pH the zeta-potential is positive which shows the potential determining role of the protonated amino groups. Because no influence of negatively charged silanol groups can be observed, it can be assumed that these groups are completely shielded by the polymer layer or involved in strong electrostatic interactions between the silica surface and the inner polymer layer.

The swelling capacity of the PVAm-*co*-PBVU layer is expected to be higher than that of the PVFA-*co*-PBVU layer as a consequence of the osmotic effect. Chloride ions, resulting from the hydrolysis reaction with hydrochloric acid, are present as counter-ions to the protonated amino groups along the polymer chain. In other words, the local salt concentration in the ionic network is remarkably high. This concentration gradient to an external aqueous phase causes osmotic pressure and more water diffuses into the polymer network where the polymer chains are increasingly stretched. The network expansion of swollen PVAm-*co*-PBVU/silica hydrogel in comparison to PVFA-*co*-PBVU/silica hydrogels in its equilibrium state has been studied by SANS (small angle neutron scattering) measurements and quantified by the correlation length (or hydrodynamic blob size) of the PVFA-*co*-PBVU/silica and PVAm-*co*-PBVU/silica network meshes [101]. The $I(q)$ curves (Fig. 16) recorded for four different temperatures (Table 2) were fitted by a sum of the Porod's law and an Ornstein–Zernicke function (Eq. 4) [102].

$$I(q) = \frac{A \cdot l}{V \cdot q^4} + \frac{I(0)_L}{1 + \xi^2 + q^2} \tag{4}$$

Table 2 Dependence on temperature of the correlation length, ξ, of PVFA-*co*-PBVU/silica and PVAm-*co*-PBVU/silica hydrogels

Temperature (°C)	ξ (nm) for PVFA-*co*-PBVU/silica	ξ (nm) for PVAm-*co*-PBVU/silica
20	2.40	7.20
25	2.99	8.99
30	3.39	6.95
40	3.26	7.16

where A is the interfacial area per volume unit V, q is the magnitude of the scattering vector, given by $4\pi/\lambda\sin(T/2)$, ξ is the correlation length, T is the scattering angle, and λ is the neutron wavelength.

The Porod behavior ($\sim q^{-4}$) at low q values arises from residual interfacial scattering from the silica surfaces. This is still present despite the solvent composition, which was chosen to exactly match the scattering density of silica by immersing the pure silica particles in different mixtures of H_2O and D_2O. We determined this matched solvent composition to have a D_2O volume fraction of 0.65. However, due to the large amount of silica in the samples (ca. 50 w/w-%) only a slightly mismatched solvent composition leads to a significant scattering from the interfaces of the silica particles.

The second term of Eq. (4) (Ornstein–Zernicke function) accounts for the liquid-like contributions from the polymer network and provides the correlation length ξ (mesh size or hydrodynamic blob size) of the PVFA-*co*-PBVU/silica and PVAm-*co*-PBVU/silica network meshes [100]. The mesh sizes for PVFA-*co*-PBVU/silica and PVAm-*co*-PBVU/silica hydrogel hybrids are listed in Table 2 as a function of temperature.

The PVAm-*co*-PBVU/silica hydrogels show a significantly larger mesh size with ξ=7.3 nm than the PVFA-*co*-PBVU/silica hydrogels with ξ=2.4 nm. This is in good agreement with the expected and macroscopically observable swelling behavior.

3.4.2
Copolymerization of Vinylsilane Functionalized Silica Particles with VFA

The copolymerization of VFA on silica surfaces requires pre-functionalization of the silica particles with a trialkoxyvinylsilane (VTS) (Scheme 8).

By means of related procedures grafting from and onto radical VFA polymerization with functionalized silica are also possible. It was found that these methods are ineffective for the synthesis of PVFA/silica hybrid materials [103]. Hence, radical copolymerization of VFA with vinylsilane-functionalized silica particles was chosen [99]. The functionalization of silica particles with VTS yields, with good reproducibility, hybrid particles (VTS-silica) with an average carbon content of 3.4 w/w-%. Co-polymerization of VFA with VTS-silica particles was performed in aqueous suspensions containing 2,2'-azobis-(2-amidinopropene) dihydrochloride (ABAC) as initiator. The

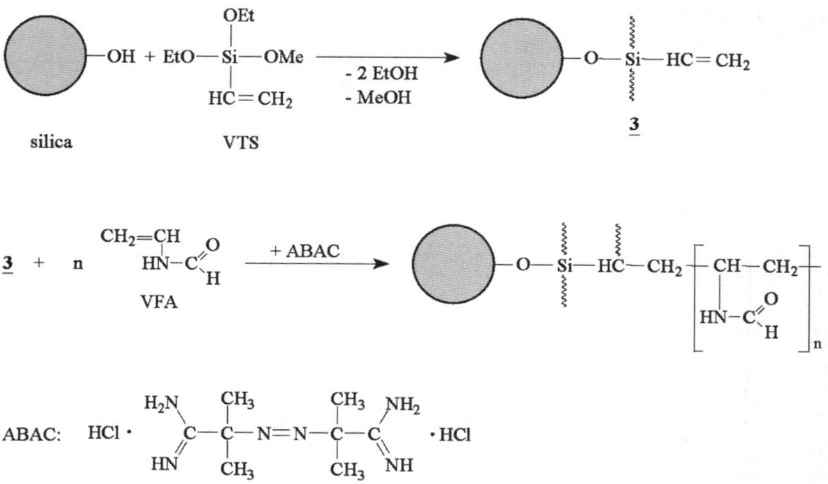

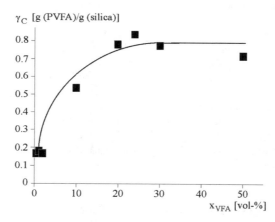

Scheme 8 Synthetic pathway to graft VFA on VTS functionalized silica surfaces

Fig. 17 Dependence of the grafted amount of PVFA (γ_C) on the initial volume fraction of VFA used to graft on vinyl silane-functionalized silica particles (x_{VFA})

PVFA obtained during the polymerization is only partially grafted on to the VTS-silica particles. The soluble non-grafted PVFA fraction was separated from the PVFA/VTS-silica hybrid particles by careful washing with water.

Figure 17 shows the amount of grafted PVFA on the VTS-silica particles as function of the initial VFA concentration (expressed as volume percent in the reaction mixture). As can be seen in Fig. 17, the largest amount of grafted PVFA can be achieved from reaction mixtures containing 24 vol-% VFA. If the volume fraction of VFA is increased a large increase in the viscosity of the reaction mixtures is observed and the suspension becomes sticky by formation of macroscopic gels. These results confirm the work of Gu et al., who reported on the radical homopolymerization of VFA in aqueous solutions [100]. In their work gel formation was observed at VFA concentrations higher than 20 w/w-% by cross-linking reactions of the formamide groups [100]. In the presence of silica it can be assumed that during gel formation diffusion of further VFA monomer to the surface is strongly restricted. Hence, at an initial VFA concentration of 20 vol-% the degree of grafted PVFA remains constant (Fig. 17).

The solid state ^{13}C CP MAS NMR spectrum of a typical PVFA/VTS-silica sample shows two main signal sets at about δ=42 ppm and δ=163 ppm, attributed to –CH–/–C H$_2$– and –C HO structure units of the PVFA (compare Fig. 14a).

As outlined previously, only the acidic hydrolysis is suitable for conversion of the formamides into amino groups. The successful conversion of the PVFA/VTS-silica hybrid particles into PVFA-*co*-PVAm/VTS-silica hybrids was confirmed by the disappearance of the signal at δ=164 ppm in the solid-state ^{13}C CP MAS NMR spectrum. The two additionally found weak signals at δ=164 ppm and δ=150 ppm result from the remaining formamide groups and carbonyl groups of formic acid and formiate ions, which can be considered as by-products of the acidic hydrolysis.

The acidic hydrolysis of the PVFA/VTS-silica hybrids is combined with loss of the grafted polymer content. It is assumed that part of the grafted PVFA chains undergo degradation, probably because of the rigorous acidic conditions.

Electrokinetic measurements of PVFA/VTS-silica and PVFA-*co*-PVAm/VTS-silica hybrids give results similar to those observed for PVFA/silica and PVFA-*co*-PVAm/silica hybrids, described previously.

4
Summary

The functionalization of spherical silica particles and silicon wafers with PVFA-*co*-PVAm has been carried out using various synthetic procedures with VFA as the key monomer. The simplest approach used to produce PVFA-*co*-PVAm/ silica hybrid materials is adsorption of PVFA-*co*-PVAm on the inorganic component from aqueous solution. Depending on the copolymer composition about 0.02 to 0.1 g PVFA-*co*-PVAm per gram silica is immobilized. The increase of the PVAm content of the co-polymer leads to a significant increase in the amount of adsorbed PVFA-*co*-PVAm.

PVFA-*co*-PVAm molecules adsorbed on silica surfaces are accessible to subsequent chemical reactions with isocyanates or fullerenes which introduce novel functionalities and irreversibly fix the PVFA-*co*-PVAm layer on the inorganic substrate.

Gold functionalized inorganic substrates are suitable for constructing patterned PVFA-*co*-PVAm functionalized surfaces, allowing incorporation of noble metals, inorganic clusters, and biologically active components.

Direct copolymerization of VFA can be carried out either with BVU, in the presence of silica particles, or with VTS, which must be pre-grafted onto silica surfaces. Both reactions can be used to immobilize large amounts of PVFA-*co*-PVAm on silica surfaces. The second described direct co-polymerization procedure is exhibited by the covalently bonded polyelectrolyte layer on the inorganic substrate. The swelling capacity of the PVFA/silica hydrogels can be controlled by the degree of hydrolysis of the PVFA.

The different syntheses used to produce stable PVFA-*co*-PVAm/inorganic oxide hybrid materials and the diversity of possible subsequent reactions which can be used to functionalize the hybrid materials make them an interesting basis for biological and technological applications.

Acknowledgement The generous financial support of this project by the Deutsche Forschungsgemeinschaft (DFG), BASF AG Ludwigshafen, and the Fonds der Chemischen Industrie is gratefully acknowledged.

We thank the following scientists for co-operation (see also the Refs. [60, 65, 69, 99]): solid-state NMR spectroscopy was performed by Dipl.-Phys. Stephanie Hesse and Professor Christian Jäger (Institute of Physics, Friedrich Schiller University Jena) and Dr Hartmut Komber (Institute of Polymer Research, Dresden); the EPR spectra were measured by Dr Manfred Friedrich (Institute of Inorganic and Analytical Chemistry, Friedrich Schiller University Jena), electrokinetic measurements were carried out by Dr Cornelia Bellmann (Institute of Polymer Research, Dresden), and SANS measurements performed in Grenoble were supported by Dr Thomas Hellweg (Iwan N. Stranski Institute, Berlin).

Furthermore, we thank Professor Christian Reichardt (University of Marburg) for generously providing betaine dyes 1 and 2, and Dr Gunnar Schornick and Dr Rainer Dyllick-Brenzinger (BASF AG, Ludwigshafen) for providing chemicals and discussions.

References

1. Mark JE, Lee CYC, Bianconi PA (1995) (eds) Hybrid inorganic–organic composites. American Chemical Society, vol 585, Washington
2. Beecroft LL, Ober OK (1997) Chem Mater 9:1302
3. Frisch HL, Mark JE (1996) Chem Mater 8:1735
4. Mann S, Burkett SL, Davis SA, Fowler CE, Mendelson NH, Sims SD, Walsh D, Whilton NT (1997) Chem Mater 11:1719
5. Iler RK (1979) The chemistry of silica. Wiley, New York
6. Scott RPW (1993) Silica gel and bonded phases. John, New York
7. Bergna HE (1994) The colloid chemistry of silica. American Chemical Society, Adv Chem Ser, Washington DC
8. Dautzenberg H, Jäger W, Kötz J, Philipp B, Seidel C, Stscherbina D (1994) Polyelectrolytes: formation, characterization and application. Hanser, München
9. Plueddeman EP (1991) Silane coupling agents. Plenum Press, New York
10. Pizzi A, Mittal KL (1994) Handbook of adhesive technology. Marcel Dekker, New York
11. Comyn J (1997) Primers and coupling agents. In: Adhesion science. Royal Society of Chemistry, Cambridge
12. Festschrift in Honor of the 75th Birthday of Plueddeman EP (1991) J Adhes Sci Technol 5:251, 425, 771
13. Heublein G, Heublein B, Hortschanski P, Meissner H, Schütz HJ (1988) Macromol Sci Chem A 25:183
14. Erler U, Heublein G, Heublein B (1990) Acta Polym 41:103
15. Köthe M, Müller M, Simon F, Komber H, Adler HJ, Jacobasch HJ (1999) Colloids Surf A 154:75

16. Sidorenko A, Zhai XW, Simon F, Pleul D, Greco A, Tsukruk VV (2002) Macromolecules 35:5131
17. Laible R, Hamann K (1980) Adv Colloid Interface Sci 13:65
18. Tsubokawa N (1992) Prog Polym Sci 17:417
19. Zaper AM, Koenig JL (1985) Polym Compos 6:156
20. Arkles B (1977) Chemtech 7:766
21. Arkles B (2000) Gelest. Gelest Inc, Tullytown, PA 19007 6308 USA, pp 16–104
22. O'Haver JH, Harwell JH, Evans LR, Waddell WH (1996) J Appl Polym Sci 59:1427
23. Ou YC, Yu ZZ, Vidal A, Donnet JB (1996) J Appl Polym Sci 59:1321
24. Johnson SA, Brigham ES, Ollivier PJ, Mallouk TE (1997) Chem Mater 9:2448
25. Sidorenko A, Minko S, Schenk-Meuser K, Duschner H, Stamm M (1999) Langmuir 15:8349
26. Minko S, Patil S, Datsuk V, Simon F, Eichorn KJ, Motornov M, Usov D, Tokarev I, Stamm M (2002) Langmuir 18:289
27. Spange S (2000) Prog Polym Sci 17:417
28. Spange S, Eismann U, Höhne S, Langhammer E (1997) Macromol Symp 126:223
29. Spange S, Höhne S, Francke V, Günther H (1999) Macromol Chem Phys 200:1054
30. Spange S, Gräser A, Müller H, Zimmermann Y, Rehak P, Jäger C, Fuess H, Baethz C (2001) Chem Mater 13:3698
31. De Campos EA, da Silva Alfaya AA, Ferrari, RT, Costa, CMM (2001) J Colloid Interface Sci 240:97
32. Decher G (1997) Science 227:1232
33. Sukhorukov GB, Donath E, Lichtenfels H, Knippel E, Knippel M, Möhwald H (1998) Colloids Surf A 137:253
34. Donath E, Sukhorukov GB, Caruso F, Davis SA, Möhwald H (1998) Angew Chem 110:2324
35. Bauer D, Killmann E, Jäger W (1998) Prog Colloid Polym Sci 109:161
36. Huguenard C, Widmaier J, Elaissari A, Pefferkorn E (1997) Macromolecules 30:1434
37. Gailliez-Degremont E, Bacquet M, Laureyns J, Morcellet M (1997) J Appl Polym Sci 65:871
38. Hao E, Wang L, Zhuang J, Yang B, Zhang X, Shen J (1999) Chem Lett 5
39. Buchhammer HM, Petzold G, Lunkwitz K (1999) Langmuir 15:4306
40. Kato T, Kawaguchi M, Takahashi A, Onabe T, Tanaka H (1999) Langmuir 15:4302
41. Lindsay GA, Roberts MJ, Chafin AP, Hollins RA, Merwin LH, Steninger JD, Smith RY, Zarras P (1999) Chem Mater 11:924
42. Roberts MJ, Lindsay GA (1998) J Am Chem Soc 120:11202
43. Shi X, Sanedrin RJ, Zhou F (2002) J Phys Chem B 106:1173
44. Lin Y, Rao AM, Sadanadan B, Kenik EA, Sun YP (2002) J Phys Chem B 106:1294
45. Zheng J, Stevenson MS, Hikida RS, van Patten PG (2002) J Phys Chem B 106:1252
46. Malynych S, Luzinov I, Chumanov G (2002) J Phys Chem B 106:1280
47. White LD, Tripp CP (2000) J Colloid Interface Sci 227:237
48. Blackledge C, McDonald JD (1999) Langmuir 15:8119
49. De Campos EA, da Silva Alfaya AA, Ferrari RT, Costa CMM (2001) J Colloid Interface Sci 240:97
50. Haupt J, Ennis J, Sevick EM (1999) Langmuir 15:3886
51. Serizawa T, Yamamoto K, Akashi M (1999) Langmuir 15:4682
52. Poptoshev E, Rutland, MW, Claesson PM (1999) Langmuir 15:7789
53. Tanaka H (1979) J Polym Sci Polym Chem Ed 17:1239
54. Achari AE, Coqueret X, Lablache-Combier A, Loucheux C (1993) Makromol Chem 194:1879
55. Dawson DJ, Gless RD, Wingward RE (1976) J Am Chem Soc 98:5996
56. Hart R (1959) Makromol Chem 32:51
57. Bayer E, Geckeler K, Weingartner K (1980) Makromol Chemie 181:585
58. Pinschmidt RK, Renz WL, Caroll WE, Yacouth K, Drescher J, Nordquist AF, Chen N (1997) J Macromol Sci Pure Appl Chem A34:1885
59. Badesso RJ, Nordquist AF, Pinschmidt RU, Sagl DJ (1996) Adv Chem Ser 248:489

60. Spange S, Madl A, Eismann U, Utecht J (1997) Macromol Rapid Commun 18:1075
61. Madl A, Spange S, Waldbach T, Anders E (1999) Macromol Chem Phys 200:1495
62. Madl A, Spange S (2000) Macromolecules 33:5325
63. Voigt I, Spange S, Simon F, Komber H, Jacobasch HJ (2000) Colloid Polym Sci 278:48
64. Voigt I, Simon F, Estel K, Spange S, Friedrich M (2001) Langmuir 17:8355
65. Meyer T, Rehak P, Jäger, C, Voigt I, Simon F, Spange S (2001) Macromol Symp 163:87
66. Eschner M, Pleul D, Spange S, Simon F (2002) In: Baselt JP, Gerlach G (eds) Dresdener Beiträge zur Sensorik, vol 16. w.e.b. Universitätsverlag, Dresden, pp 149–153
67. Shi J, Seliskar CJ (1997) Chem Mater 9:821
68. Tamaki R, Chujo Y (1999) Chem Mater 11:1719
69. Madl A, Spange S (2000) Macromol Symp 161:149
70. Roth I, Spange S (2001) Macromol Rapid Commun 22:1288
71. Meyer T, Spange S, Simon F (2003) J Colloid Interface Sci, in press
72. Meyer T, Hellweg T, Spange S, Jäger C, Hesse S, Bellmann C (2002) J Polym Sci A, 40:3144
73. Voigt I, Simon F, Estel K, Spange S (2001) Langmuir 17:3080
74. Roth I, Seifert A, Hartmann P, Spange S, in published results
75. Bauer D, Killmann E (1997) Macromol Symp 126:173
76. Rehmet R, Killmann E (1999) Colloid Surf A 149:323
77. Van de Steeg HGM (1992) Langmuir 8:2538
78. Durand G, Lafuma F, Audebert R (1988) Prog Colloid Polym Sci 76:278
79. Wang T, Audebert R (1988) Colloid Interface Sci 121:32
80. Davies RJ, Dix L, Toprakcioglu C (1989) Colloid Interface Sci 129:145
81. Reichardt C (1994) Chem Rev 94:2319
82. Reichardt C, Harbusch-Görnert E (1983) Liebigs Ann Chem 721
83. Spange S, Reuter A, Lubda D (1999) Langmuir 15:2103
84. Spange S, Reuter A (1999) Langmuir 15:141
85. Spange S, Reuter A, Prause S, Bellmann C (2000) J Adhes Sci Technol 14:399
86. Jensen WB (1991) In: Mittal KL, Anderson HR (eds) Acid–base-interactions. VSP, Utrecht
87. Allemand PM, Khemani KC, Koch A, Wudl F, Holzer K, Donovan S, Gruner G, Thomson JD (1991) Science 253:301
88. Sijbesma R, Srdanow G, Wudl F, Castro JA, Wilkins C, Freedman SH, DeCamp DL, Kenyon GL (1993) J Am Chem Soc 115:6510
89. Tabata Y, Ikada Y (1999) Pure Appl Chem 71:2047
90. Yamakoshi YN, Yagami T, Fukuhara F, Sueyoshi S, Miyat NJ (1994) Chem Soc Chem Commun 517
91. Chen XL, Jenekhe SA (1999) Langmuir 15:8007
92. Hungerbuchler H, Guldi DM, Asmus KD (1993) J Am Chem Soc 115: 3386
93. Beeby A, Eastoe J, Heenan RK (1994) J Chem Soc Chem Commun 173
94. Feng W, Miller B. (1999) Langmuir 15:3152
95. Reed CA, Bolskar RD (2000) Chem Rev 100:1075
96. Mrzel A, Mertelj A, Omerzu A, Copic M, Mihailovic D (1999) J Phys Chem B 103:11256
97. Force field calculations were done with PCMODEL 4.0 (Serena Software, Bloomington, IN, USA) using the MMX-force field
98. Hollemann AE, Wiberg E (1976) Lehrbuch der anorganischen Chemie, 81–90 Aufl. Berlin, de Gruyter, p 801
99. Meyer T, Spange S, Hesse S, Jäger C, Bollmann, C (2003) Macromol Chem Phys, 204:725
100. Gu L, Zhu S, Hrymak AN, Pelton RH (2000) Macromol Rapid Commun 22:212
101. Higgins JS, Benoit HC (1994) Polymers and neutron scattering. Clarendon Press, Oxford
102. De Gennes PG (1979) Scaling concepts in polymer physics. Cornell University Press, London
103. Meyer T (2001) PhD Thesis, Chemnitz University of Technology

Received October 2002

Adv Polym Sci (2004) 165:79–150
DOI 10.1007/b11268

Polyelectrolyte Brushes

Jürgen Rühe[1] · Matthias Ballauff[2] · Markus Biesalski[1] · Peter Dziezok[3] ·
Franziska Gröhn[4] · Diethelm Johannsmann[5] · Nikolay Houbenov[6] ·
Norbert Hugenberg[7] · Rupert Konradi[1] · Sergiy Minko[6] · Michail Motornov[6] ·
Roland R. Netz[8] · Manfred Schmidt[4] · Christian Seidel[9] · Manfred Stamm[6] ·
Tim Stephan[5] · Denys Usov[6] · Haining Zhang[1]

[1] Institut für Mikrosystemtechnik, Albert-Ludwigs-Universität, 79110 Freiburg, Germany
  E-mail: ruehe@imtek.uni-freiburg.de
  E-mail: biesalsk@imtek.uni-freiburg.de
  E-mail: konradi@imtek.uni-freiburg.de
  E-mail: hzhang@imtek.uni-freiburg.de
[2] Polymerinstitut, Universität Karlsruhe, 76128 Karlsruhe, Germany
  E-mail: Matthias.Ballauff@chemie.uni-karlsruhe.de
[3] Fa. Proctor and Gamble, 65823 Schwalbach, Germany
  E-mail: dziezok.p@pg.com
[4] Institut für Physikalische Chemie, Johannes Gutenberg-Universität, 55099 Mainz,
  Germany
  E-mail: groehn@mpip-mainz.mpg.de
  E-mail: mschmidt@mail.uni-mainz.de
[5] Institut für Physikalische Chemie, TU Clausthal, 38678 Clausthal Zellerfeld, Germany
  E-mail: johannsmann@pc.tu-clausthal.de
  E-mail: stephan@mail.uni-mainz.de
[6] Institut für Polymerforschung Dresden, 01069 Dresden, Germany
  E-mail: houbenov@ipfdd.de
  E-mail: minko@ipfdd.de
  E-mail: motornov@ipfdd.de
  E-mail: stamm@ipfdd.de
[7] Fa. Zeiss, Curfeßstrasse 9, 73430 Aalen, Germany
  E-mail: nhugenberg@aol.com
[8] Sektion Physik, Ludwig-Maximilians-Universität, 80333 München, Germany
  E-mail: netz@theorie.physik.uni-muenchen.de
[9] Max-Planck-Institut für Kolloid- und Grenzflächenforschung, 14476 Golm, Germany
  E-mail: seidel@mpikg-golm.mpg.de

Abstract Polyelectrolyte brushes constitute a new class of material which has recently received considerable interest. The strong segment–segment repulsions and the electrostatic interactions present in such systems bring about completely new physical properties of such monolayers compared to those consisting of either non-stretched or non-charged polymer chains. In this review some recent progress on the theory, synthesis, and swelling behavior of polyelectrolyte brush systems in different environments is discussed. The height of the polyelectrolyte brushes is studied as a function of the molecular weight and graft density on both planar and spherical surfaces. In addition it is elucidated how the brushes are affected by external conditions such as the ionic strength of the surrounding medium, the presence of multivalent or polymeric ions and in some cases by the pH of the contacting solution. Two more specific cases, the synthesis and characterization of mixed polyelectrolyte brushes and cylindrical polyelectrolyte brushes, in which charged polymer chains are attached to the backbone of other polymers, are described in more detail.

1
Introduction

Polymer brushes are systems in which chains of polymer molecules are attached through an anchor group to a surface, to an interface, or to a backbone of other polymer molecules in such a way that the graft density of the polymers is high enough that the attached chains are stretched away from the surface in a brush-like conformation (Fig. 1) [1–5]. Chain stretching starts, essentially, when the chains at the substrate surface begin to overlap and increases strongly with increasing graft density of the chains. The stretching process is the result of segment–segment interactions within the attached chains and is counterbalanced by the elastic free energy of the chains, which opposes stretching, as any chain stretching leads to a strong loss of conformational entropy. This balance between repulsion and rubber-like elasticity of the chains leads to a new equilibrium at a higher energetic state than that of isolated coils.

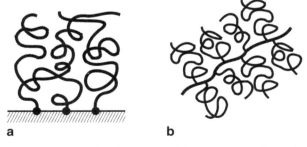

a   **b**

Fig. 1 Schematic illustration of brush regimes: (**a**) at a solid surfaces: (**b**) of chains attached to the backbone of another polymer

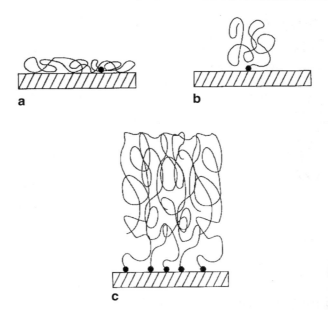

Fig. 2 Schematic illustration of the conformation of polymers end-attached to surface: (a) "pancakes"; (b) "mushrooms"; (c) "brushes"

In pioneering research in the late 1970s Alexander [6] and de Gennes [7, 8] described the scaling of neutral polymeric monolayers that are irreversibly attached by one end to a surface. Depending on the grafting density, which is defined as the inverse distance of two neighboring surface-attached polymer molecules, three different regimes are distinguished (Fig. 2). In the first two essentially single chains are attached to the surface, i.e. the distance between anchoring points is larger than the size of the molecules and the surface-attached chains do not overlap. If the polymer segments now have a strong tendency to be adsorbed by the surface the macromolecules typically have a flat, "pancake-like", conformation at the surface. In contrast to this, if "non-adsorbing" polymers are attached to the surface in such a low grafting density regime, a "mushroom"-like conformation can be observed. Here largely unperturbed polymer coils are grafted to the surface. Finally, if the macromolecules are attached with a high density to the surface, chain crowding leads to a stretching of the chains, normal to the surface, and the surface-attached polymer layer assumes a "brush"-like conformation. The theory of polymer brushes in general at both planar and curved surfaces is today quite well developed both analytically and numerically and the basic features of such systems in respect of brush height and segments density profile as a function of graft density, molecular weight, and solvent quality are rather well understood (at least for monodisperse systems).

Polymer brush systems are theoretically and experimentally of great interest as the change in the conformational entropy of the attached chains and the concomitant stretching of the polymer chains bestow interesting new properties on such materials. Examples of systems with new and enhanced properties are surfaces with ultralow friction obtained by coating of two surfaces with polymer brushes and sliding then against each other [8a] or the so-called autophobic behavior, in which materials coated with surface-attached polymer chains do not become wetted by free polymer even if the surface-attached and the free chains are chemically identical [8b].

The physical properties of charged macromolecules in contact with a solid surface are fundamentally different from that of similar layers consisting of uncharged polymers [9]. In addition to segment–segment interactions and changes in conformational entropy, which are present in neutral polymer films, electrostatic interactions come into play and the structure and properties of polyelectrolyte layers are almost exclusively dominated by these interactions. Mutual repulsion between the charged polymer segments and electrostatic forces between the polyelectrolyte molecules and the surface (especially if the latter is also charged), strongly influence the strength of interaction with the substrate and the physical properties of the layers [5]. Furthermore the conformation of the polymer chains in the brush depends strongly on external conditions, especially the ionic strength of the surrounding medium, the presence of multivalent ions, and, if weak polyelectrolytes are studied, on the pH value of the contacting aqueous phase.

In the following text recent progress in the field of polyelectrolyte brushes is reviewed. In the first section the theory of polyelectrolyte brushes is briefly summarized and some more recent simulation results are shown. In the next section several aspects of the synthesis and the elucidation of the physicochemical properties of polyelectrolyte brushes are discussed. In the following section adaptive and responsive polymer surfaces based on mixed polyelectrolyte brushes are introduced, followed by a section on cylindrical polyelectrolyte brushes, in which charged polymer chains are attached to the backbone of other polymers. Finally some perspectives for further developments in the field of polyelectrolyte brushes are given.

2
Theory of Polyelectrolyte Brushes: From Scaling to Molecular Simulations

2.1
Neutral Polymer Brushes

If neutral and charged polymer brushes are exposed to solvents, a very interesting and rich phase behavior can be observed. As a consequence, such surface-attached brushes have become the focus of considerable theoretical efforts in studies on the structure and phase behavior of such polymer chains in contact with a solvent. The thickness of a neutral brush scales linearly with the degree of polymerization, N, which is in stark contrast to the well-known characteristics of free polymer chains in a good solvent, where the radius of the coils scales as $R \propto N^{0.58}$. For neutral brushes the simple scaling laws connecting grafting density and molecular weight with brush height were first derived by Alexander [6] and de Gennes [7, 8]. These scaling laws have since largely been confirmed by more sophisticated theories [10–15], direct numerical solutions [5, 16], simulations [17], and experiments [18–20]. The stretching of a surface-attached macromolecule relies on a balance between the elastic energy and the osmotic pressure. Briefly, individual chains can be taken as entropic springs with a spring constant $\kappa=3kT/(Na^2)$ where kT is the thermal energy, N is the chain length and a is the segment length. An ensemble of such entropic springs exerts a pressure of $P_{el}=\rho_a\kappa L=3\rho_a kTL/(Na^2)$. Here ρ_a is the grafting density, and L is the height of the swollen surface-attached brush. With v being the excluded volume and setting c the segment concentration to $c=N\rho_a/L$, the osmotic pressure can be written as $P_{osm}=1/2vc^2kT$. Finally, the following scaling relation is obtained by equating the elastic and the osmotic pressure:

$$L \propto N \cdot a \cdot \rho_a^{1/3} \tag{1}$$

A more detailed overview of theoretical studies on the structure of neutral polymer brushes is beyond the scope of this article and can be found in several review articles [12, 21, 22].

2.2
"Strong" Polyelectrolyte Brushes

The swelling behavior of a charged polymer brush fundamentally differs from that of an otherwise very similar, but uncharged brush. Conformational changes of the surface-attached macromolecules are now mainly governed by electrostatic interactions and the osmotic pressure of the counter-ions, rather than the osmotic pressure of the macromolecular segments. Understanding the sometimes peculiar behavior of such systems, an important distinction has to be made regarding "strong" and "weak" polyelectrolytes. For strong polyelectrolyte brushes the number and position of charges on the chain is fixed. Variation of the pH or the ion-concentration will not affect the number of charges. Because, in this case, charges are permanently associated with a certain chemical group, brushes of strong polyelectrolytes are often called "quenched brushes". The scaling behavior of strong polyelectrolyte brushes has been studied extensively from a theoretical point of view [23–29]. In weak polyelectrolytes a dissociation–association equilibrium exists. Only the average degree of charging is given by the equilibrium constant, which is governed by the pH of the solution, but not the position of the individual charges. As the charges are mobile within the brush they are sometimes termed "annealed brushes". Such brushes can show behavior quite different from that known for the quenched ones [30–35].

Depending on grafting density, ρ_a, degree of dissociation, f, and on the ionic strength of the medium quenched polyelectrolyte brushes exhibit a wide spectrum of different structures. In strongly charged brushes with a large effective charge density of grafted polyelectrolytes, i.e. when both grafting density and charge fraction are sufficiently large, effectively all the counter-ions are trapped inside the brush. This is the case when the Gouy–Chapman length $\lambda_{GC}=(2\pi l_B fN\rho a)^{-1}$, which is the height at which counter-ions are effectively bound to a surface of charge density $ef N\rho_a$, is smaller than the brush height, L [23]. Here, l_B is the Bjerrum length and e is the elementary charge. Provided that Coulomb correlations are not important and the counter-ion distribution is homogeneous in the lateral directions, in this regime the brush height scales linearly with the degree of polymerization N, but it is independent of the grafting density ρ_a. The fact that the brush thickness does not depend of the number of attached chains per surface area can be directly seen from simple scaling arguments. In this case, which is called the osmotic brush regime, the brush height is determined by the balance between the osmotic pressure of the counter-ions and the restoring force of the stretched chains. Because both osmotic pressure of the counter-ions and elastic pressure of the chains depend linearly on grafting density, the "osmotic brush" thickness becomes independent of ρ_a [23, 24]

$$L_{OsB} \propto N \cdot a \cdot f^{1/2} \tag{2}$$

If strong polyelectrolyte brushes with very high grafting densities are considered the excluded volume interaction can no longer be neglected. If steric effects dominate over electrostatic effects the elastic energy is again balanced by a second virial term and the behavior of the neutral brush, discussed above in Eq. (1), is recovered. Precisely speaking, the prefactor is increased due to an electrostatic contribution to the excluded volume. In this case, however, the scaling properties of the brush depend strongly on the relative strength of the interaction of the polymer with the aqueous environment compared to the polymer–polymer interactions. The interactions of the polymer with water, however, will be in many systems strongly altered by charge recombination.

On the other hand, if the grafting density and/or the degree of dissociation is reduced the Gouy–Chapman length becomes larger than the brush height and the counter-ion cloud extends far beyond the rim of the brush. In this case the osmotic pressure of the counter-ions becomes irrelevant against uncompensated electrostatic repulsion between charged monomers. This results in the charged or Pincus brush regime, where the elastic and the bare electrostatic pressures are balanced, giving a brush height which is dependent on the grafting density [23]

$$L_{PB} \propto N^3 a \cdot f^2 l_B \rho_a \tag{3}$$

However, as the phase regime where such interactions dominate is rather narrow and the absolute changes in brush height are expected to be quite small, an experimental confirmation of such a polyelectrolyte brush phase remains extremely difficult. Solving the non-linear Poisson–Boltzmann equation without any pre-assumption about the counter-ion distribution perpendicular to the grafting surface, Zhulina and Borisov have shown that in the large intermediate region the brush height grows monotonically with grafting density when the counter-ion distribution extends beyond the rim of the brush [29]. In the two limits $\lambda_{GC} < L_{OsB}$ the osmotic brush regime and for $\lambda_{GC} > L_{OsB}$ the Pincus brush are reproduced.

The picture of the polyelectrolyte brush behavior changes if salt is added to the solution. Pincus describes the swelling behavior of strong polyelectrolyte brushes in solutions of varying ionic strength [23, 24]. At low external salt concentrations c_S screening only takes place at the outer edge of the brush leaving the overall brush height unaffected. Once the concentration of added ions in solution reaches the concentration of the free counter-ions inside the brush, the screening associated with the external salt reduces the osmotic pressure of the mobile ions inside the brush. The osmotic pressure forcing the chains to stretch from the surface at high ionic strengths can be derived as $p \approx c^2 kT/2\, c_S$, where c is the segment density. Note that in this case the osmotic pressure depends quadratically on the density of the polymer segments, in analogy with phase regimes where excluded volume interactions dominate the behavior of surface-attached chains. The effective excluded volume parameter v_{ef}, however, now scales as $v_{ef} \propto c_S^{-1}$. Again, the osmotic pressure balances the entropic elasticity. As a consequence, the height L_{Salt} of the polyelectrolyte brush at high external salt concentrations becomes a function of the external salt concentration and of the grafting density [23]

$$ L_{Salt} \propto Na \cdot c_S^{-1/3} \cdot \rho_a^{-1/3} \tag{4} $$

It can be seen that on addition of salt the brush height decreases by a weak power law, $L_{Salt} \propto c_S^{-1/3}$. This behavior is often called "salted brush". Here the surface-linked monolayer behaves like a neutral brush with an enlarged (electrostatic) excluded volume. Note that the grafting density has a strong influence on the amount of free mobile ions per chain inside the brush. Therefore, the height of such polyelectrolyte brushes is expected to become less sensitive to the external salt concentration if chains at high graft density are considered.

2.3
"Weak" Polyelectrolyte Brushes

Weak polyelectrolyte brushes consist of surface-attached polymer chains, where an equilibrium exists between neutral, undissociated and the charged, dissociated moieties. In such systems the degree of dissociation depends on the local pH value [30–32]. Examples of such systems are weak polyacids. At low pH a large abundance of protons will result in a low charge density due to protonation of the salt moieties. A number of molecular parameters such as the charge density on the brush, the concentration of free counter-ions, and the degree of swelling can therefore be tuned via adjustment of the pH. Upon the addition of large amounts of salt, weak polyelectrolyte brushes shrink similar to what is expected for strong polyelectrolyte brushes (Eq. 4) [33–35].

Theoretical expectations for the swelling behavior of weak brushes at low salt concentrations are, on the other hand, somewhat counter-intuitive. If only small amounts of salt are added to the solution the brush height increases. The reason for such a behavior is that the local concentration of protons in the brush is governed by the requirement of charge neutrality. However, when the ambient solvent contains ions other than protons, some of these cations can be exchanged with the protons without violation of charge neutrality. Hence, the degree of dissociation of the acid/base moieties on the polymer chains changes. Generally, some of the cations might also recombine with the acidic groups to yield a salt. The binding constant for this kind of association, however, is much lower than the binding constant of the pure acid/base equilibrium.   As a consequence, a net increase of charge therefore remains, resulting in an increase in osmotic pressure. Thus the height of the weak polyacid brush increases with increasing salt concentration. This process has been thoroughly studied by Zhulina and coworkers [34, 35] and by Fleer [32]. The outcome is a weak power law dependency predicted for the scaling of the brush height as a function of the external salt concentration and the grafting density of the surface-attached chains:

$$L \propto Na \cdot c_S^{1/3} \cdot \rho_a^{-1/3} \tag{5}$$

However, it should be noted that this is only a somewhat simplified summary of the theory of PEL brushes. For a more detailed discussion the reader is referred to Refs. [6–35].

2.4
Simulations

2.4.1
Introductory Remarks

Simulations are promising tools for checking theoretical models and probing quantities and regimes which are not easily observable experimentally. However, despite strong efforts in recent years [36], simulations of polyelectrolytes still remain challenging. Because of the special methods required for treating the long-range Coulomb interaction they are computationally rather expensive. For this reason few simulation studies of polyelectrolyte brushes have been reported in the literature [37–39].

2.4.2
Simulation Model and Method

In work by the Seidel group a freely jointed bead–spring model was used where the polyelectrolytes consist of charged monomers (for partially charged chains also uncharged species) which are connected by non-linear springs (FENE potential) and grafted by one end to an uncharged surface (Figure 3). Counter-ions are treated as free charged particles, while solvent

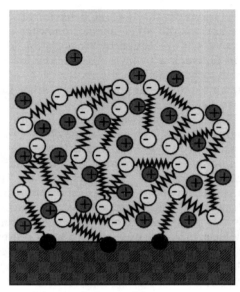

Fig. 3 Schematic picture of the simulation model

molecules are replaced by a dielectric background and a heat bath. The
chains are assumed to be in a good solvent modeled by a purely repulsive
Lennard–Jones potential. All particles except the anchor segments interact
repulsively with the grafting surface at short distances and all charged enti-
ties interact with the bare Coulomb potential. Thus, the total potential consists of four contributions $U_{tot}=U_{FENE}+U_{LJ}+U_{wall}+U_{Coulomb}$. A schematic picture of the simulation model is shown in Fig. 3. Equilibrium properties are
studied by stochastic molecular dynamics. The equation of motion for particle i at position $\mathbf{r}_i(t)$ is the Langevin equation

$$m\frac{d^2\mathbf{r}_i}{dt^2} = -\nabla_i U_{tot} - m\Gamma\frac{d\mathbf{r}_i}{dt} + \mathbf{W}_i(t) \tag{6}$$

where all particles are assumed to carry the same mass m. Γ is a friction
constant which couples the particles to a heat bath. The system is held at
thermal equilibrium by a Gaussian random force $\mathbf{W}_i(t)$. A major task in any
simulation of charged systems is the correct treatment of the long-range interaction U_{Coul}. To calculate Coulomb forces and energies in the case of a
2D+1 slab geometry, we use a technique proposed by Lekner [40] and modified by Sperb [41].

2.4.3
Simulation Results

Varying the Bjerrum length l_B, we obtain a non-monotonic behavior of the brush height with a maximum at rather small coupling strengths (see Fig. 4) [42]. Together with the average height of the chain ends $\langle z_e \rangle$ we plot the Gouy–Chapman length λ_{GC} (above). For our choice of parameters the average bond length is $a=0.98\ \sigma$, with σ being the Lennard–Jones radius.

For simplicity, henceforth we use $a=\sigma$ and restrict the discussion to completely charged systems ($f=1$). Considering Fig. 4 we address three points:

1. At $l_B>0.5a$ the stretching of the chains is smaller than in a corresponding uncharged brush, indicating the influence of attraction due to electrostatic correlations.
2. The maximum brush height occurs at $l_B \approx 0.1\ a$. For smaller l_B, the Gouy–Chapman length becomes larger than the brush height. Hence, counter-ions are free to leave the brush giving rise to its relaxation back to the reduced extension of a quasi-neutral brush.
3. Close to the maximum height we obtain a stretching of the chains up to about 2/3 of their contour length. This is certainly out of the range where Gaussian elasticity can be applied. For $l_B=0.1\ a$, Fig. 5 shows snapshots from the equilibrium trajectories.

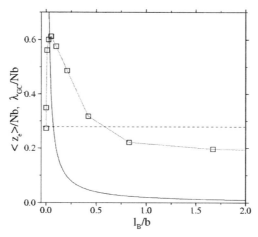

Fig. 4 Brush height versus Bjerrum length: average end-point height (*squares*) and Gouy–Chapman length (*thick line*). The *dashed line* gives $\langle z_e \rangle$ of a corresponding uncharged brush

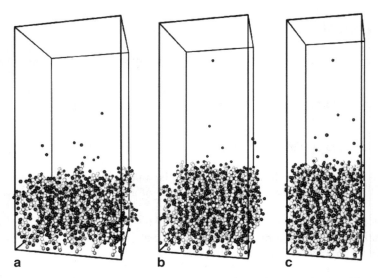

Fig. 5 Completely charged brushes ($N=30$) at grafting density of (**a**) $\rho_a b^2 = 0.04$, (**b**) $\rho_a b^2 = 0.06$, (**c**) $\rho_a b^2 = 0.09$, and (**d**) $\rho_a b^2 = 0.12$. The Bjerrum length is $l_B = 0.1\ b$

Note the following points:

1. Immediately one can realize that the chains are strongly stretched and become aligned perpendicular to the grafting surface.
2. Obviously there occurs only a weak dependence of the stretching on grafting density.
3. Almost all counter-ions remain trapped inside the brush.

In Fig. 6 the average thickness of the brush and of the counter-ion layer, measured by the first moments of the corresponding density profiles, is plotted as a function of grafting density. As indicated by the snapshots shown in Fig. 5, at $f=1$ almost all the counter-ions remain captured inside the brush. Changing l_B from 0.7 a to 0.1 a the relation between $\langle z_m \rangle$ and $\langle z_{ci} \rangle$ remains almost unaffected while the dependence of the stretching on ρ_a is drastically changed. In contrast to the well-known scaling law for charged brushes in the osmotic regime, which predicts a thickness independent of the anchoring density, in any case we observe a non-negligible effect. At $l_B = 0.1a$ the dependence becomes rather weak with a power-law exponent $\alpha = 1/5$ [43]. At stronger interaction, i.e. at $l_B = 0.7\ a$, we obtain a new collapsed regime where the monomer density becomes independent of the grafting density, resulting in a linear scaling of the brush height with ρ_a [44].

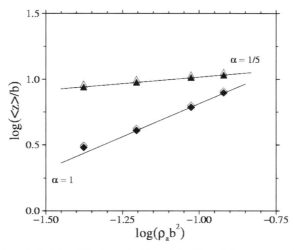

Fig. 6 Average brush height (*filled symbols*) and height of the counter-ion layer (*open symbols*) of completely charged brushes versus grafting density. l_B=0.7 b (*diamonds*) and l_B=0.1 b (*triangles*). The lines give power laws $\langle z_e \rangle \sim \rho_a{}^{\alpha}$

2.5
Extended Theory of Polyelectrolyte Brushes

2.5.1
Coulomb Correlations and Collapsed Brush Regime

Including electrostatic correlations which cause an attractive interaction, the nature of the novel collapsed brush regime (CB) can be understood within an extended scaling model [45]. The simplest way to treat correlations between mobile charges is within the Debye–Hückel approximation where the corresponding free energy (per unit area) reads $F_{DH} \sim -L \, (Nf l_B \rho a / \, L)^{3/2}$. The resulting phase diagram in the $(f\rho_a)$-plane is plotted in Fig. 7. For weak electrostatic coupling $l_B < v_2{}^{1/3}$ (v_2 is the second virial), we observe the diagram of states obtained by Borisov et al. [28]. For strong coupling $l_B > v_2{}^{1/3}$, however, the novel collapsed regime can occur at strong charging. In this phase, the equilibrium brush height $L_{CB} \sim N\rho_a v_2{}^2 f^{-3} l_B{}^{-3}$ results from a competition between steric repulsion and attractive Coulomb correlations. For the particular choice of simulation parameters, at l_B=0.7 a we have $l_B{}^3 \approx v_2$ which corresponds to the cross-over between strong and weak electrostatic coupling. In this case, the scaling theory allows no quantitative comparison, but it correctly describes the trends:

1. with increasing f one enters the CB regime [45] (not shown here); and
2. the CB regime disappears when l_B is reduced.

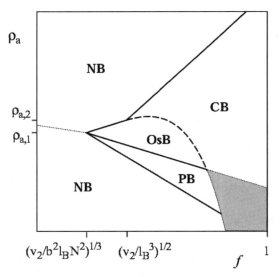

Fig. 7 Scaling theory phase diagram at $v_2/l_B^3 < 1$ in logarithmic scales with quasi-neutral (NB), osmotic (OsB), Pincus (PB) and collapsed (CB) brush regimes

2.5.2
Lateral Inhomogeneity and Non-Linear Osmotic Brush Regime

In all previous theoretical studies of polyelectrolyte brushes it was commonly assumed that counter-ions are distributed uniformly in the lateral directions parallel to the grafting plane. However, inhomogeneous distributions were obtained both in simulation [44] and experiment [46]. Using non-linear Poisson–Boltzmann theory, lateral inhomogeneity is explicitly taken into account in Ref. [47]. It was shown there that lateral inhomogeneity of the counter-ion distribution yields a weak dependence of brush height on anchoring density, although the counter-ions are assumed to be trapped inside the brush. The corresponding brush regime is called non-linear osmotic. Representing a stretched polyelectrolyte chain, which is supposed to consist of spherical monomers, by a cylindrical rod, there are different possibilities to choose its diameter. One way is to relate it directly to the bead size (Fig. 8a, left-handed). Another possibility is to calculate it under the condition that the polymer volume remains constant (Fig. 8a, right-handed). Similarly, there are different possibilities to match the cylindrical cell model with the rectangular simulation box (Fig. 8b). Clearly the different ways to set the inner and outer radii of the cell model influence the translational entropy of mobile counter-ions. To compare theoretical predictions quantitatively with simulation data and with experimental results one has to take into account the finite extensibility of polyelectrolyte chains. Although it does not change the scaling behavior of brush height with grafting density [48], at large stretching it strongly influences the absolute value of chain

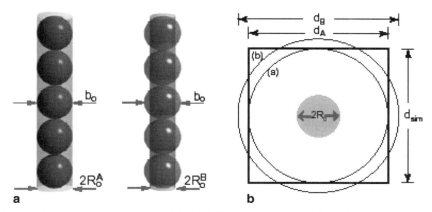

Fig. 8 Poisson–Boltzmann cell model: (**a**) different ways to model a polyelectrolyte chain by a cylindrical rod, (**b**) different ways to choose the cell boundaries

extension. In fact, including lateral inhomogeneity of the counter-ion distribution together with non-linear elasticity of the polyelectrolyte chains provides reasonable agreement between simulation data and theoretical predictions without any fitting parameter [47]. In Fig. 9 we compare the simulation result and theoretical predictions for the two different models discussed above. Within the order of the systematic error of the model we obtain reasonable agreement with the simulation data. Interestingly, there is some indication of weak dependence of chain stretching on anchoring density from recent experiments with polyelectrolyte brushes [49].

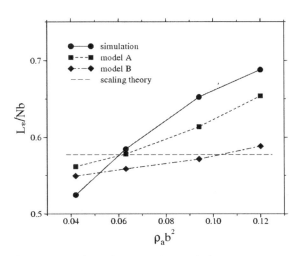

Fig. 9 Brush height versus grafting density at $l_B=0.1$ b, $f=1$. Simulation results and theoretical predictions

3
Synthesis and Physicochemical Characterization of Polyelectrolyte Brushes

3.1
Introductory Remarks

Polyelectrolyte brushes at solid surfaces can be formed either by adsorption of charged block copolymers from solution or by anchoring amphiphilic, charged block copolymers at the air–water interface, compression, and subsequent transfer of the layer to a solid substrate [9, 21, 50]. A third very important alternative is chemical attachment of charged polymers to the surface of the substrate. In the following sections we will focus primarily on the latter, which in one way or the other includes the formation of a covalent chemical bond between the charged polymer molecules and the solid to which the polymer chains are attached. In general, there are two well-established routes for terminal attachment of polymer molecules to a solid surface, one in which macromolecules containing reactive end-groups react with appropriate reactive sites at the surface of the substrate, commonly called a "grafting to" procedure, the other where the monolayer is formed directly at the surface using a monolayer of surface-attached initiator molecules. The initiator monolayer is then used to start a polymerization reaction. In the latter process the polymer chains are grown directly on the surface of the substrate in situ and such a process is frequently termed "surface initiated polymerization" or "grafting from".

3.2
Polymer Brushes via Chemisorption

The "grafting to" process uses specifically (ex-situ) synthesized end-functionalized macromolecules that are chemically attached to appropriate reactive sites on the substrate. Essentially such a process is an extension of the formation of self-assembled monolayers (SAMs) of low molecular weight compounds at solid surfaces to their high molecular weight analogs. Accordingly polymer molecules with chlorosilane and alkoxysilane end groups have been prepared for immobilization on glass and silicon surfaces, and polymers with thiol and disulfide groups for attachment to gold surfaces. Although most of the work published so far has been devoted to the formation of "neutral" polymer brushes there are also some examples where charged polymer brushes have been established by this method.

In one example, polystyrene sulfonate (PSS) brushes terminated with trichlorosilane anchor groups are attached both to spherical [51] and to planar silicon oxide surfaces [52–54]. To obtain such structures in the first step a monolayer consisting of neutral polystyrene molecules is formed by reaction of the silane anchor groups with silanol groups on the silicon/silica substrates. The neutral polymer is then transformed into the polyelectrolyte in a second step via a polymer-analogous sulfonation reaction. The thus ob-

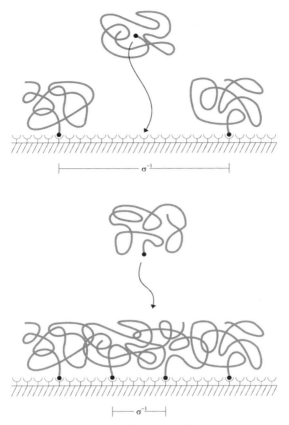

Fig. 10 General scheme of a "grafting to" process for the end-attachment of polymer molecules to solid surfaces. The early build-up of the polymer monolayer is governed mainly by the diffusion of the molecules to the surface. However, at later stages adsorbed amount and thickness are limited due to kinetic and thermodynamic hindrances

tained monolayer consist of surface-attached polymer molecules which have a high molecular weight ($N{\approx}1900$), low polydispersity ($M_n/M_w<1.10$) and a degree of sulfonation of roughly $f{\approx}0.6$ [54]. The value of the latter parameter was obtained from FTIR spectra of the surface-attached monolayers. The chemisorption of polymers from solution is a very simple method of generating polymer monolayers, but also has some limitations, which cannot be overcome easily. One serious limitation is that only relatively simple polymer molecules such as polystyrene can be surface-attached by this method, as the high reactivity of the anchor groups prevents the coexistence of most functional groups in the polymer. For example chlorosilane end groups would not tolerate any amino, hydroxyl, acid, and a variety of other functional groups. In addition to this the thicknesses of the surface-attached monolayers obtained by this method are limited quite strongly [55, 56]. At the beginning of the surface-reaction, when the surface is still more or less unoccupied, free chains can easily diffuse to the surface and react with their counterparts on the surface. Once the surface starts to become significantly covered with already attached polymer chains, however, the diffusion of further chains against the concentration gradient into the film becomes very slow (Fig. 10). This diffusion barrier leads to a very strong kinetic hindrance of the attachment of further chains and limits the film thicknesses accessible in experimental time frames to just a few nanometers. In addition to this the macromolecules already attached to the surface have to stretch in order to accommodate further chains to the surface. Accordingly, the attachment of further molecules is energetically quite costly and even if the kinetic limitations for film formation could be circumvented, the graft densities which can be obtained by this technique will still be limited for thermodynamic reasons [57]. Thus polymer monolayers obtained by the "grafting-to" route are usually only in the "mushroom" or at the low graft density end of the "brush" regime and the chain stretching is typically rather weak.

As a consequence of the general chemistry of such a process polyelectrolyte brushes prepared by this technique always require a two-step approach where a neutral brush system is formed in a first step and then charged sites are introduced in a second step. However, one should be aware that such a post-synthetic modification process might lead to additional complications if the transformation is not quantitative or if side reactions occur. For a detailed understanding of the structural behavior of such brushes a homogeneous polymer-analogous transformation throughout the whole brush is a basic requirement.

3.3
Polymer Brushes via Surface-Initiated Polymerization

An alternative, which enables some of the problems of the "grafting to" processes to be overcome, is the formation of polymer molecules directly at the surface of the substrate by using a monolayer of surface-attached initiators and starting a surface-initiated polymerization reaction as shown schematically in Fig. 11 [2, 57–59]. In this case only monomer has to diffuse to the growing chain end, thus eliminating the problem with the diffusion barrier, which strongly limits film growth in the "grafting to" case. As essentially any chain reaction for the formation of polymer molecules will give the desired structures, a number of different initiator systems and a variety of polymerization techniques have been employed. For example, techniques such as radical chain [58–60], living cationic [57, 61], and controlled living (atomic transfer reaction and tempo-mediated) [62] polymerization have been used. Substrate materials include spherical [58, 59] and planar [60, 62–64] silicon oxide surfaces, spherical [61] and planar metal surfaces [57], and soft polymeric surfaces [65]; a large number of functional monomers have been used so far [2].

Rühe and coworkers used surface-attached monolayers of an azo initiator to start the free radical polymerization of a variety of different monomers.

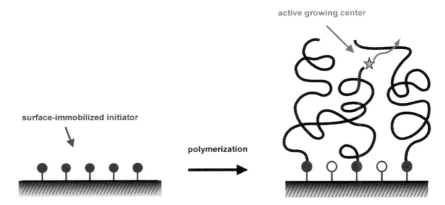

Fig. 11 Schematic description of a "grafting from" process for the in-situ generation of surface-attached functional polymer monolayers

Neutral [2, 58–60] and charged polymer brushes [2, 63, 64] with high molecular masses and high grafting densities of the surface-bound macromolecules have been established. While molecular weights of the surface-attached polymer chains of more than 10^6 g mol^{-1} could be realized, the distances of the surface-attached chains were in some cases less than 3 nm.

In general, the (dry) thickness of polymer brushes is a function of the molecular mass of the attached polymer chains and of the grafting density [2, 59]. For conventional free radical chain polymerization processes, choosing an appropriate monomer/solvent ratio in the polymerization step controls the molecular mass of the attached chains. The grafting density of the surface-attached chains can be controlled by adjusting the conversion of the initiator, i.e. by choosing the polymerization time. Typically, polymer brushes having a wide range of thicknesses starting from a few nanometers to more than 1000 nm in the dry, solvent free state have been reported [2, 59, 60, 63]. Further parameters that influence the molecular weight and/or the graft density of the surface-attached chains are the temperature during the polymerization process, as this determines the rate of dissociation of the initiator and thus the radical concentration in the film, the radical efficiency, and transfer reactions. Brushes prepared by this free radical polymerization method typically have a polydispersity of the order of $M_n/M_w \approx 2$ (at least if the conversion of the monomer is kept at a reasonably low level), due to the nature of the radical polymerization mechanism and the associated termination steps [58].

In a first example, for a densely grafted PEL brush system positively charged quaternized poly-4-vinylpyridine brushes have been prepared by following a two step approach [2, 63, 64, 66]. In the first step a neutral poly-4-vinylpyridine monolayer is prepared and, subsequently, charges are introduced by a second, polymer-analogous quaternization step. The grafting density of the parental neutral brush is adjusted by varying the polymerization time (Fig. 12) [63, 64]. The substrates were planar silicon substrates in

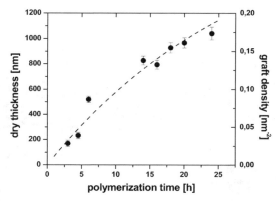

Fig. 12 Dry thickness (solvent-free state) of a neutral poly-4-vinylpyridine brush attached to a planar solid surface as a function of polymerization time. The *solid line* represents the theoretically calculated thickness from the polymerization kinetics. For more details see text

most cases. The yield of the polymer-analogous reaction was analyzed by using FTIR and measurement of the increase in the mass of the polymer film during the quaternization reaction (Fig. 13), as the addition of the alkyl halogen compound leads to a significant mass increase of the repeat units of the polymer. The conversion of the polymer-analogous reaction was found to be quantitative, within the experimental error of the surface-analytical techniques employed, ensuring homogeneous distribution of the charged sites throughout the brush [63]. Polyelectrolyte brushes are prepared with high molecular masses of the attached chains ($M_n > 10^6$) and high grafting densities (i.e. low distances d between neighboring chains, $d \approx 2-5$ nm) [66]. However, one step syntheses of polyelectrolyte brushes were also reported–by using the same immobilized azo-initiator monolayer and surface-initiated polymerization monolayers of polystyrenesulfonate or methacrylic acid were generated at the surface in situ. Again the grafting density and the molar mass of the polyelectrolyte chains were controlled by adjusting the polymerization parameters.

An interesting aspect of this way of preparing polyelectrolyte brushes is that it is also very simple to prepare (statistical) copolymer brushes. To this only the two monomers have to mixed in the appropriate ratio and (if the copolymerization parameters permit) statistical copolymer brushes can be synthesized by the surface-initiated polymerization route. This allows for some interesting structural variations of the polyelectrolyte brushes. For example, if the second comonomer is not charged, but also hydrophilic, a copolymer brush is formed, which can be strongly swollen by water, but where the average degree of charging can be strongly varied all the way from almost uncharged copolymers to the fully charged polyelectrolytes.

In an example by the Ballauff group, which follows essentially a very similar strategy but with a different synthetic approach, latex particles with a

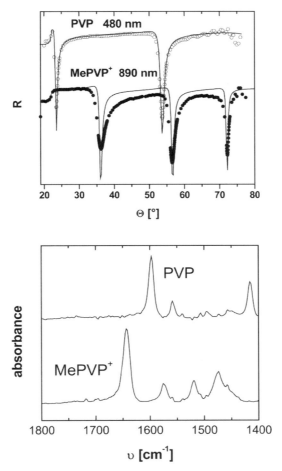

Fig. 13 (*above*) Waveguide-spectra (p-polarization) of a 490 nm thick PVP brush on a LaSFN9/Au/SiO$_x$ substrate, and the same sample at a constant relative humidity of 70% after quaternization with methyl iodide (MePVP, thickness: 870 nm); (*below*) FTIR detail spectrum of a PVP brush attached to both sides of a silicon wafer, and the same sample after the polymer-analogous quaternization

polystyrene core and a shell of polystyrene sulfonate or polyacrylic acid chains (Fig. 14) were synthesized by a photo-emulsion polymerization process [2, 65]. In a first step polystyrene particles with typical sizes between 50 and 100 nm and narrow polydispersity were prepared by conventional emulsion polymerization. In a second step the latices were covered in situ with a polymeric layer, where the polymer molecules contain side-groups with a photo-initiator (seeded emulsion polymerization). This second emulsion polymerization is carried out under starved conditions to avoid formation of new particles and to ensure a defined core shell morphology. In the third

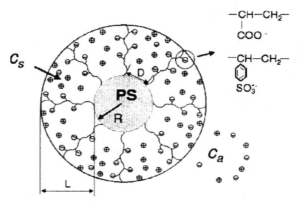

Fig. 14 Schematic representation of a spherical polyelectrolyte brush prepared by the photoemulsion grafting from technique (details see text). The brush consists of a solid polystyrene core and surface-attached strong (PSS) or weak (PAA) polyelectrolyte brush shell (Reprinted from Ref. [71] with permission from the American Physical Society)

and final step the polyelectrolyte brushes were generated on the surface of the particles by photoinitiation using the surface-bound initiators in the presence of monomer. To carry out this reaction the latex particles were dispersed in monomer solution and irradiated in the reaction mixture. Although the latex with the surface-attached initiator presents a turbid system, the strong light scattering in such systems is practically completely elastic, so that photoinitiation can be used for such heterophase systems with satisfactory efficiency. Sufficient dilution increases the interparticle distances and ensures that no recombination of radicals growing on two different particles occurs. This is important because such a process would ultimately lead to cross-linking of the complete system. The brush height, i.e. the hydrodynamic thickness of the surface-attached layers, was determined by dynamic light scattering. The brush height was found to increase with increasing monomer concentration during the polymerization process. The results obtained show that the photo-emulsion polymerization technique offers a further convenient means of preparation of stable polyelectrolyte-modified particles.

3.4
Structural Characterization of Polyelectrolyte Brushes at Solid Surfaces

3.4.1
Swelling of Polyelectrolyte Brushes in Humid Air

For a variety of applications, especially in the life sciences, coatings applied to the surface of a given material with the aim of making it more biocompatible must be stable in aqueous buffer solution and must have good interfacial properties for biomolecules. The stability of polymer films used in any of such life science application in contact with a gaseous, liquid or more complex biological environment is of great importance. If the polymer layers are swollen when they are exposed to a certain environment, mechanical stress due to the swelling process can cause the film to become completely or partially detached from the surface (delamination). Apart from the integrity of the films another serious issue is the displacement of the polymer molecules of the coating by other competing adsorbents from the environment. The latter aspect is especially important in bio-oriented applications as the coatings are exposed to rather complicated environments and some molecules competing for surface sites have a very strong tendency to adsorb and displace the molecules which are already adsorbed on the surface.

One example of a practical application where polyelectrolytes are of crucial interest are immunoassays, where charged polymers are attached to surfaces and are then exposed to protein solution with the aim of loading of the polymer layer with a reproducible amount of these proteins. The protein-loaded particles or planar films thus obtained are in turn exposed to analyte solutions containing other proteins. If the proteins in the film match the proteins in solution, protein–protein complexes are formed, which are then visualized and/or quantified. In order to complete such a process successfully knowledge about the swelling of the polyelectrolyte layer in buffer solutions and the interaction of such a layer with proteins has to be established. Further questions, which are of great importance for such polyelectrolyte systems are the behavior of the monolayers in contact with common impurities present in contacting solutions especially traces of multivalent ions or tensides.

The first and simplest example of such a swelling process is when polyelectrolyte brushes are exposed to humid air. As the films interact strongly with water they swell to some extent on exposure to ambient humidity. The extent of the swelling depends of course strongly on the structure of the polymer. Knowledge about the swelling in humid air is important as most of the methods for determining the thickness of the layer or the mass of the attached polymer are carried out under ambient conditions and thus do not give only the mass of the polymer chains, but include the amount of incorporated water. If this is not considered, a significant error concerning the amount of attached polymer results.

In an attempt to evaluate the significance of this parameter a monolayer of poly- N -methyl-(4-vinylpyridinium) iodide, synthesized by quaternization of poly-4-vinylpyridine with methyl iodide, was brought into contact with different controlled environments using a cell which was held at constant temperature and constant relative humidity. The humidity inside the cell was set to a chosen value by placing a small vessel containing a saturated aqueous salt solution in it and sealing it carefully. To generate environments with varying humidity, a number of different saturated aqueous salt solutions were chosen. As the different salts have each an individual saturation concentration due the solubility of the individual salts in water at the given temperature, also the equilibrium water vapor pressure of these salt solutions varies. As in all cases saturated solutions are chosen, the water vapor

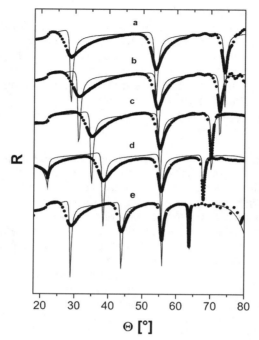

Fig. 15 Waveguide spectra (p-poL) of a 740 nm thick (dry thickness) MePVP brush covalently attached to the surface of a LaSFN9/Au/SiO$_2$ substrate. The spectra were measured in different (constant) rel. humidities a = 0%, b = 35%, c = 70%, d = 87% and e = 100% rel. humidity at 25°C. The *solid lines* are the calculated reflection curves obtained from Fresnel modeling

pressure and thus the humidity inside the cell is highly reproducible. To obtain "dry" film thicknesses KOH pellets were introduced into the cell, as the moisture content of air in contact with KOH is below 10^{-5} mbar.

The film thicknesses (and refractive indices) of the layers on top of 90° prisms were measured by waveguide spectroscopy, which records the reflectivity of the sample as a function of the angle of incidence. Experimental details have been reported elsewhere [64]. Reflectivity curves and Fresnel calculations (superimposed on the experimental curves as solid lines) are shown in Fig. 15. A large uptake of water vapor molecules corresponding to an increase in thickness of about 150% (Fig. 16) is observed if the film environment is changed from 0 to 100% relative humidity at ambient temperature. The increase of the film thickness is consistent with a concurrent decrease of the refractive index of the film [64]. This strong influence of the humidity of the environment on the film thickness shows that it is indeed important to take such a process into account and that all measurements where the film thicknesses of such polyelectrolyte layers are recorded do not give the mass of the attached polymer directly but have to be significantly

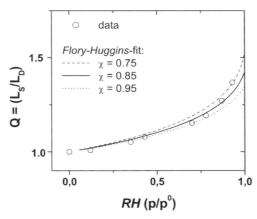

Fig. 16 Degree of swelling of a 740 nm thick (dry thickness) MePVP brush covalently attached to the surface of a LaSFN9/Au/SiO$_x$ substrate as a function of the relative humidity (*RH*) of the environment. For more details see text

corrected to allow for the air humidity-induced swelling. If the swelling of the film due to the humidity content of the surrounding air is neglected the actual mass of the polymer in the film will be quite significantly overestimated in such measurements.

Although it is intuitively quite clear that the film thickness of the surface-attached polyelectrolyte layers increases with increasing moisture content of the air, it is interesting to analyze this behavior in more detail. The experimental data shown in Fig. 16 can be well described by the Flory–Huggins model, if a χ-parameter of $0.75 < \chi < 0.95$ is used for the model-fit (the fits are shown superimposed on the experimental data as solid lines in the figure). The high value for the interaction parameter clearly indicates that humid air is a "bad" solvent for the surface-attached polyelectrolyte molecules. During the swelling process in humid air the uptake of water vapor molecules is not sufficient to enable significant dissociation of the polyions and the counterions. As a consequence the hydrophobicity of the backbone determines the properties and the interaction of the water vapor molecules with each other is strong compared to the interaction with the surface-attached polymer segments. Thus it is not surprising that for the swelling of a polyelectrolyte brush in humid air large values of the χ-parameter are observed.

3.4.2
Swelling Behavior of "Strong" Polyelectrolyte Brushes

Much more interesting, however, than the swelling in humid air is the behavior of polyelectrolyte brushes in contact with water or salt solutions. The first experimental study of the swelling properties of a surface-attached polyelectrolyte monolayer obtained by a "grafting to" process has been reported by Auroy and coworkers [51]. Polystyrene sulfonate layers chemically

attached to a high surface-area silica gel are prepared by grafting polystyrene molecules, by means of chlorosilyl end-groups, on to the solid surface,
and, subsequently, introducing the charges via a polymer-analogous sulfonation reaction. The silica gels thus obtained were then studied with neutron
scattering techniques. However, due to problems during the synthesis, such
as chain degrafting and possible side-reactions, no systematic investigations
of the influence of the different parameters on the brush height could be performed.

Transferring this preparation method in later studies on to planar solid
surfaces the same group published more systematic structure analysis on
the phase behavior of polystyrene sulfonate brushes [52–54]. In this case,
the height and segment density of the polyelectrolyte brushes were investigated with neutron reflectometry. The surface-attached charged macromolecules exhibit a strong stretching normal to the surface, due to the electrostatic repulsion of the charged segments. The authors claim that some of
the chains could even be "fully" stretched. As predicted by mean-field theory the average swollen height of the polyelectrolyte brushes scales almost
linearly with the length of the attached polymer and is furthermore independent of the grafting density [53]. Adding small electrolytes to the solution
the brush height is affected above a certain concentration, and, due to electrostatic screening of the repulsive forces inside the layer, shrinks with a
weak power-law as predicted by theory.

In another example, the swelling behavior of both surface-attached neutral poly-4-vinylpyridine (PVP) and positively charged poly-N-methyl-(4-
vinylpyridinium) iodide (MePVP) brushes prepared by the "grafting from"
method were studied under "good solvent" conditions by means of multiple-
angle null-ellipsometry [67, 68]. In such experiments the brush is grown on
the base of a prism, which is used to couple in the light. The ellipsometric
parameters ψ and Δ are then recorded as a function of the angle of incidence. From these spectra the refractive index profile and hence the segment
density profile can be reconstructed either by fitting prespecified functions
to the data or by performing a direct Fourier transformation [68]. This technique allows the segment-density profile of the brushes synthesized on the
high refractive index glass prisms to be obtained in a rather simple way. The
charged polymer brush was generated from the neutral precursor by a polymer-analogous quaternization reaction in the same way as described above.
The brushes investigated consist of surface-attached macromolecules with
high molecular masses ($M_n>10^6$ g mol^{-1}) and high grafting densities
($10^{-2}<\sigma<1$ chains per nm^2).

Because the MePVP brushes are directly related to the PVP brushes from
which they are synthesized, comparison of the brush height of the neutral
and charged systems allows direct visualization of the consequence of the
"switching on" of electrostatic interactions. A schematic depiction of the situation and a direct comparison of the segment density profiles for a swollen
brush in the neutral state (PVP) and after addition of the charges (MePVP)
is shown in Fig. 17. In both cases a good solvent (PVP/methanol and
MePVP/water, "salt free") for the respective brush is employed. The segment

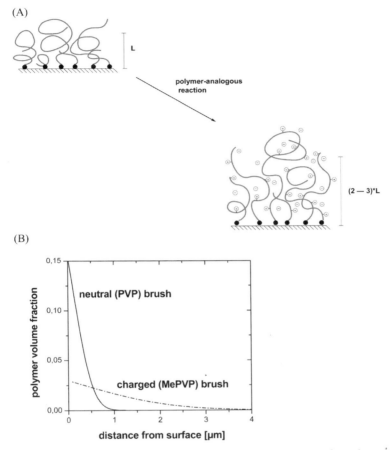

Fig. 17 (a) Schematic depiction of the increase in swollen thickness of a polymer brush in a good solvent after conversion of a neutral brush into the charged species by introduction of electrostatic forces. (b) Segment density profiles of a neutral PVP and a charged MePVP brush. Both samples have an identical number of segments per surface-attached polymer molecule similar grafting densities and are both in contact with a good solvent (methanol for PVP, water for MePVP)

density profiles of both the neutral and charged brushes do not show the expected parabolic shape as predicted by mean-field theory and simulation, but rather follow an exponential function (Fig. 17b). Although this discrepancy is not yet well understood, one possible explanation concerns with the polydispersity of the experimental system. As a free radical chain polymerization reaction was used to grow the polymer chains from the surface, the polydispersity of the surface-attached molecules is significant (an estimate gave $M_n/M_w \approx 2$), whereas the theory was developed for monodisperse systems. It has been shown theoretically for neutral polymer brushes, by Milner

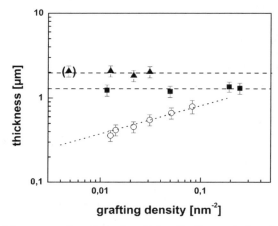

Fig. 18 Brush thickness as a function of graft density for neutral and charged brushes. The experimental results are shown together with the scaling laws predicted by mean-field theory (lines in the figure: $L \sim \sigma^{0.33}$ for the neutral brush, and $L \sim \sigma^0$ for the polyelectrolyte brush)

et al., that the polydispersity directly impacts on the segment density distribution [68a]. For polyelectrolyte brush systems the influence of polydispersity on the segment density distribution is yet to be investigated, both experimentally and theoretically. The brush height in these and all the following ellipsometric experiments is defined as twice the first moment of the segment density distribution.

In agreement with mean-field theory, the swollen brush height of the PVP brushes increases with graft density ($L \propto \sigma^{1/3}$) whereas the height of the highly charged polyelectrolyte brush in pure, "*salt-free*" water is almost independent of the graft density ($L \propto \sigma^0$). In Fig. 18 the results of these studies are shown together with the theoretically expected power-laws, which are superimposed on the experimental data in Fig. 18. In the case described here, the surface-attached neutral polymer chains stretch to about 40% and the positively charged species to about 65% with respect to the theoretically calculated contour-length [66, 69].

Apart from the effect of the external salt concentration on the static behavior of such systems, the dynamic properties of such brushes are also of great importance. Gelbert et al. [70] investigated the influence of added electrolytes on the internal structure of a polystyrenesulfonate brush prepared by the "grafting from" method by means of noise analysis of a scanning force microscope (SFM) cantilever. The dynamics of the surface-attached polyelectrolyte brush, in particular fluctuations of the brush thickness under compression and varying ionic strength of the liquid phase, were determined by bringing the brush into contact with a cantilever of a scanning force microscope. To do this a silica sphere with a diameter of 10 µm was glued to the tip of the cantilever to obtain well-defined geometry in the in-

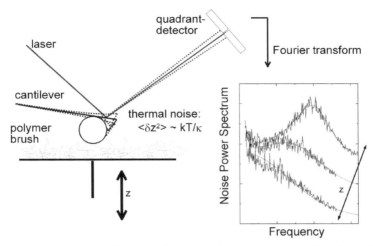

Fig. 19 Experimental set-up for thermal noise analysis experiments. The dynamic information is contained in the random motion of a sphere which is in contact with the brush and suspended on the cantilever of a scanning-force microscope (SFM)

teraction zone. The schematics of the set-up are shown in Fig. 19. The thermally activated motion of the cantilever is modified by the interaction between the brush and the tip and can be analyzed through a noise power spectral density [70]. The outcome of such a noise analysis is, in principle, equivalent to a force modulation spectrum.

The polyelectrolyte brush shrinks strongly on addition of electrolytes. At low or moderately low salt concentrations (c_s=0.01 mol L^{-1}) the force profiles resemble those of a soft brush. At salt concentrations of c_s= 0.03 mol L^{-1}, however, the profile of the static force resembles more closely that of a hard surface. Interestingly, if the behavior of the PEL brush is studied close to the collapse point significantly increased compressibility can be observed. However, the compressibilty shows no bistability, which indicates that the transition between the brush and the collapsed state is not a true first-order transition, although this would be expected from mean-field theory. One possible explanation of this behavior would be that the polydispersity of the surface-attached chains smoothens the transition.

In another example the swelling behavior of a spherical brush consisting of a solid polystyrene core and a polystyrene sulfonate shell was investigated with respect to the ionic strength and the pH of the liquid phase [71]. The brushes were grown on the surface of the latex particles by photoemulsion polymerization, which has been described above. The height of the brush layer was determined by dynamic light-scattering experiments. It is evident, that the swelling behavior of such a system must be considerably different from that of the same system on a planar surface, as with increasing distance from the surface the accessible volume increases with the third power of the

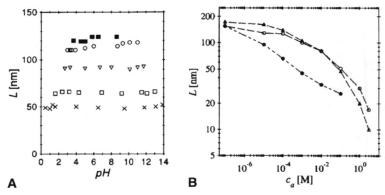

Fig. 20 (a) Brush thickness as a function of pH of the medium for a surface-attached spherical PSS brush at different constant potassium chloride concentrations (*crosses*, 1 mol L^{-1}; *open squares*, 0.1 mol L^{-1}; *triangles*, 0.01 mol L^{-1}; *open circles*, 0.001 mol L^{-1}; *filled squares*, 0.0001 mol L^{-1}). The radius of the polystyrene core is $R=68$ nm; the contour length of the attached chains is $L_c=147$ nm, the grafting density is $\sigma=0.037$ nm^{-2}. (b) Brush thickness as a function of potassium chloride concentration of the medium for a surface-attached spherical PSS brush at constant pH=7 (*triangles*, $R=72$ nm, $L_c=159$ nm, $\sigma=0.027$ nm^{-2}; *open circles*, $R=54$ nm, $L_c=141$ nm, $\sigma=0.031$ nm^{-2}; *filled circles*, $R=54$ nm, $L_c=141$ nm, $\sigma=0.031$ nm^{-2} in MgSO$_4$). (Reprinted from Ref. [71] with permission from the American Physical Society)

radius of the shell, which of course influences the segment density profile strongly.

Figure 20 shows the dependence of the brush thickness L on the pH (A) and salt concentration (B) of the liquid phase. As expected from mean-field theory, the height of the strong polyelectrolyte brush at various grafting densities and molecular masses of the surface-attached macromolecules is independent of pH at a given constant electrolyte concentration of the medium so that the PSS chains are fully dissociated under all circumstances [71].

The height of such spherical PEL brushes as a function of the concentration of added monovalent salts was investigated at constant solution pH using two different PSS brushes. The data (Fig. 20b) show that at low ionic strength the PSS brush height is almost independent of the external salt concentration. At higher salt concentrations the polyelectrolyte brush shrinks. However, it is evident that this behavior cannot be described by simple scaling laws. The discrepancy between the experimental results and scaling law prediction are most likely not caused by the polydispersity of the attached macromolecules, but rather is determined by the fact that the contour-length of the polymers is of the order of all other length-scales of the system (e.g. the radius of the particle), and the brush height becomes very sensitive to these pertinent parameters [71]. For a more detailed insight into the scaling of spherical polyelectrolyte brushes the reader is referred to discussions in Ref. [71] and references therein.

3.4.3
Swelling of "Weak" Polyelectrolyte Brushes

An interesting, and with regard to basic physics somewhat different, case is present if the surface-attached polyelectrolyte chains consist of a weak polyelectrolyte. The local charge density of a weak polyacid or polybase is not fixed but is a function of the local concentration of protons inside the brush. If the pH is raised or lowered new charges are generated on the brush or disappear due to charge recombination. However, also within the brush the situation is not same in all different locations. Since the degree of dissociation of weak polyelectrolytes depends on the polymer concentration and accordingly the segment density in the brush, which is a function of the z -coordinate, also the charge density varies within the brush as function of the distance from the substrate.

In one example polymer segment density profiles of weak polyacid brushes consisting of polymethacrylic acid (PMAA) chains were investigated as a function of the pH of the environment by means of multiple-angle ellipsometry [68]. The polymer brushes were prepared by surface-initiated polymerization of methylacrylic acid as described above.

Figure 21 shows the segment density profiles of such PMAA brushes as obtained by Fresnel-modeling, and, more directly (i.e. "model free"), by FT analysis of ellipsometric spectra at different solution pH [68]. Figure 22 shows the swollen brush height L obtained from these experiments.

It is evident that the behavior of a polyelectrolyte brush consisting of surface-attached weak polyacid chains is markedly different from that of a strong polyelectrolyte brush described above, as the number of charges on the polymer chain varies strongly with changing pH. At very low pH all sites are protonated and the brush carries no net charge. Increasing the pH generates more charges inside the brush, which in turn causes an increase in electrostatic repulsion and, as a result, the polymer chains are stretched out

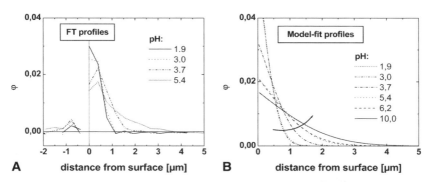

Fig. 21 Profiles of a surface-attached weak polyacid brush as obtained from FT analysis of ellipsometric data at different pH. The pH is given in the figure. Analysis of the same ellipsometric data by Fresnel modeling. Best fits to complementary error functions at different pH given in the figure. For more details see text

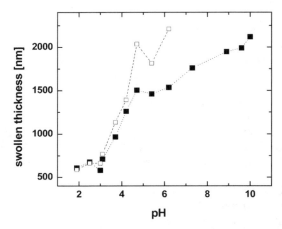

Fig. 22 Swollen thickness of a polymethacrylic acid brush attached to a planar lanthanum oxide surface as a function of the pH of the solution. The thickness is determined by FT analysis (*closed symbols*) and by Fresnel modeling (*open symbols*) of ellipsometric data. For details see text

further into solution. Accordingly the overall brush height increases. It seems clear that outer regions of the brush are more affected because the segment density decreases with increasing distance from the surface. At high pH a highly extended tail is observed that is also expected from SCF calculations on such systems. However, theoretical calculations only treat strictly monodisperse systems, and theoretical calculations for neutral surface-attached chains of the influence of polydispersity on the segment density profile also yield profiles that have long tails, which extend far into the solution. Consequently, it is not yet clear what part of the observed behavior is due to the change in the local pH and which is due to the polydispersity of the surface-attached chains.

Dynamic properties of such PMAA polyelectrolyte brushes, investigated by SFM noise analysis, are reported in a further communication [70]. Noise analysis reveals that the PMAA brush shrinks with decreasing solution pH, which is consistent with the findings of the optical measurements for the same system described above. Analysis of static force profiles shows that the grafted polyelectrolyte brushes are softest at the highest degree of swelling (Fig. 23) and that the brushes appear to have extended outer tails, which is consistent with both the ellipsometric analysis and mean-field predictions. From force–distance profiles it is concluded that the polyelectrolyte brush collapses at 5<pH<6, which, again, is consistent with the ellipsometric results on the same system (Fig. 23). In a manner very similar to that for strong polyelectrolyte brushes a highly compressible region at near-collapse condition is found for weak PMAA brushes [70].

A qualitative similar picture is observed for brushes consisting of the same kind of polymer chain attached to a solid spherical substrate. By anal-

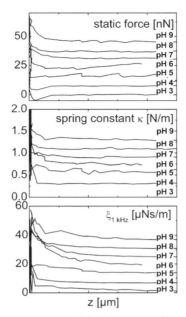

Fig. 23 Viscoelastic parameters measured during approach to a PMAA brush under various conditions of pH. The friction coefficient is increased at intermediate pH

ogy to the corresponding strong polyelectrolytes such spherical PAA brushes have recently been prepared by the photoemulsion polymerization method [71–73].

The influence of the pH and ionic strength of the surrounding medium on the brush height was determined by dynamic light scattering. Figure 24 shows the brush height as a function of the external pH. The brush height follows essentially the same behavior as that seen by ellipsometric measurements of PMAA brushes on flat surfaces. In agreement with expectations the stretching of the chains becomes more pronounced when a low ionic strength of the medium is considered, because at high ionic strength screening of the charges starts to set in Ref. [70].

When such a spherical weak polyelectrolyte brush is studied, a somewhat counterintuitive behavior can be observed, which has been predicted by mean-field theory, and is sometimes called "anomalous salt effect". While at high salt concentrations the brush height shrinks in a way very similar to that observed with strong polyelectrolytes, due to charge screening, at low ionic strength of the medium the height of a weak poly acid brush increases with increasing salt concentration as shown in Figure 25 for a PMAA brush in contact with sodium nitrate solutions. Thus at intermediate concentration a maximum of the brush height as a function of the external salt concentration is observed (Fig. 25). While at first view it looks somewhat peculiar that the brush starts to swell with increasing concentration, it becomes directly

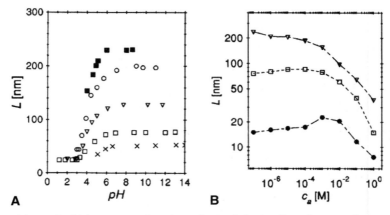

Fig. 24 (a) Brush thickness as a function of pH of the medium for a surface-attached spherical PAA brush at different constant potassium chloride concentrations (*crosses*, 1 mol L^{-1}; *open squares*, 0.1 mol L^{-1}; *triangles*, 0.01 mol L^{-1}; *open circles*, 0.001 mol L^{-1}; *filled squares*, 0.0001 mol L^{-1}). The radius of the polystyrene core is: R=66 nm; the contour length of the attached chains is: L_c=228 nm; the grafting density is: σ=0.039 nm^{-2}; (b) Brush thickness as a function of the potassium chloride concentrations of the medium for a surface-attached spherical PAA brush at constant pH=7 (*triangles*, R=66 nm, L_c=228 nm, σ =0.039 nm^{-2}; *open squares*, R=105 nm, L_c=123 nm, σ=0.018 nm^{-2}; *filled circles*, R=57 nm, L_c=42 nm, σ=0.038 nm^{-2} in MgSO$_4$). (Reprinted from Ref. [71] with permission from the American Physical Society)

clear if not only the screening of charges but also the dissociation equilibrium is considered. With increasing concentration of monovalent ions the protons of the carboxylic acid become more and more replaced. As generally sodium salts of carboxylic acids are more strongly dissociated than their acid counterpart the polymer chain becomes more and more charged and the osmotic pressure inside the brush starts to increase. However, when the salt concentration is further increased, this effect becomes overcompensated by charge screening and the brush shrinks again (see also the discussion in Ref. [71]).

3.5
Interaction of Brushes with Multivalent Ions

When the brushes are brought into contact with solutions containing multivalent ions the picture described above changes dramatically. To demonstrate the effect the same PMAA brushes as described above were exposed to solutions containing multivalent ions. Figure 25 shows as examples changes of brush thickness with increasing ion concentration when the brushes are exposed to sodium, calcium, and aluminium nitrate solutions [74].

The brush thickness passes through a maximum only with the monovalent sodium ion. If the brush is exposed to solutions containing bivalent alkaline earth metal ions such as calcium, the brush collapses at intermediate

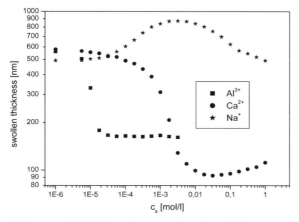

Fig. 25 Swollen thickness of PMAA brushes as a function of the external salt concentration for aqueous sodium, calcium, and aluminium nitrate solutions. Values at a concentration of 10^{-6} mol L^{-1} correspond to the swelling in pure MilliQ water. The dry film thicknesses were 42 nm (*stars*), 45 nm (*circles*) and 46 nm (*squares*)

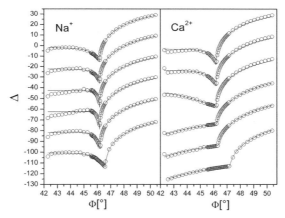

Fig. 26 Representative ellipsometric spectra (Δ as a function of the incidence angle; Ψ is omitted for clarity) of PMAA brushes on LaSFN9 prisms swollen in aqueous solutions of $NaNO_3$ and $Ca(NO_3)_2$. The concentrations are 10^{-5}, 10^{-4}, 10^{-3}, 10^{-2}, 10^{-1} and 10^0 mol L^{-1} from the *top* downwards and the offset in Δ by which the curves are shifted for clarity is 20. The *solid lines* represent model calculations using a complementary error function to describe the segment density profile

concentrations (around 10^{-3} mol L^{-1}; Fig. 25) and exhibits no maximum in brush thickness. Figure 26 compares typical ellipsometric spectra (Δ only) for a PMAA brush immersed into aqueous sodium and calcium solutions of different concentration. Even simple visual inspection of the spectra reveals a remarkably different behavior for the monovalent and bivalent ions and

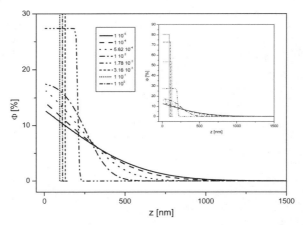

Fig. 27 Representative segment density profiles of the collapse of a PMAA brush on a LaSFN9 prism swollen in aqueous solutions of Ca(NO$_3$)$_2$ having different external salt contents as denoted in the figure. The profiles were obtained from model calculations based on the ellipsometric spectra shown in Fig. 26

allows qualitative conclusions to be drawn. For calcium as the external salt the flattening of the curve which connects the angle of incidence with the ellipsometric parameter Δ below the critical angle indicates that the brush height is becoming dramatically smaller and the layer is getting closer to a uniform box-like profile. Figure 27 shows some representative segment density profiles for the same PMAA brush at different calcium concentrations obtained from model fits to the ellipsometric spectra in Fig. 26. The transition from a fuzzy profile to a box like profile at about 10^{-3} mol L^{-1} can be seen quite clearly.

It is, furthermore, quite interesting to see that well above the collapse concentration the brush seems to re-swell slightly upon addition of more calcium ions to the solution. Although the effect is rather small and within the experimental error, this feature seems to occur in all samples. One possible explanation for this behavior is that with increasing calcium concentration equilibrium formation requires that more and more calcium ions bind only one carboxylic acid group within the film, leaving a number of moieties with a net positive charge behind. Due to the increasing amount of charge on the polymer chains they become more soluble and the film thickness increases again.

Last, PMAA brushes in contact with aluminium solutions show similar behavior to the calcium case; however, the collapse concentration is found to be lower by roughly two orders of magnitude. If an aluminium salt solution is added to the brush the collapse starts at concentrations of $<10^{-5}$ mol L^{-1} (Fig. 25). Furthermore, it is interesting to note that the film still contains rather large amounts of water in the collapsed state and remains much more swollen, compared to the same film which has been collapsed by exposure to calcium.

Theoretical studies that deal with the thickness trends of weak polyacid brushes show that if ionic interactions alone are responsible for swelling behavior the brush will display a maximum in its thickness with increasing ionic strength. This behavior is predicted for both monovalent and multivalent ions. In the described studies, however, only the monovalent sodium salt case is in agreement with this theory [75]. As already described above the thickness increases with increasing salt concentration, at sufficiently low salt concentrations, in the so-called osmotic brush regime, since the addition of salt facilitates the dissociation of the acid groups and leads to an increase in the degree of ionization. Above a critical ionic strength, however, the brush thickness decreases with increasing salt concentration, due to the increased screening of the charged groups. This is called the salted brush regime.

That completely different behavior is observed for bivalent calcium and trivalent aluminium counter-ions can be attributed to the fact that ionic interactions alone do not govern the brush structure and that the situation is somewhat more complex. Indeed, there have been experimental investigations on the interaction of weak polyacids in different topologies with multivalent cations in the literature [76–78]. Osmotic de-swelling measurements on sodium polyacrylate gels [77], and combined static and dynamic light scattering measurements on sodium polyacrylate chains [78], show a volume transition at specific ion concentrations. In addition, bivalent transition metals and trivalent cations lower the elastic free energy by forming cross-links between the polymer chains. How exactly the collapse of the brushes proceeds will depend on subtleties of the interaction of the multivalent ion with the polymer backbone, especially the strength of the formed complex and its interaction with water. It can be envisioned that, if the complexation constant is becoming too high, i.e. the complex becomes too stable, no re-equilibration of the brush structure becomes possible.

3.6
Interaction of Polyelectrolyte Brushes with Other Polyelectrolytes

The following describes preliminary results on the formation of monomolecular layers of surface-attached polyelectrolyte–polyelectrolyte complexes by using polyelectrolyte brushes as substrates [79]. In particular, the differences between strong and weak polyelectrolyte systems were studied. The surface-attached polyelectrolyte complexes obtained were used for further layer-by-layer build-up of polyelectrolyte multilayers. Subsequent processes in which the substrates were alternately dipped into a strong and a weak polyelectrolyte will be termed strong/weak systems. The corresponding systems consisting of two varieties of strong PEL molecules will be called strong/strong.

In the first step the influence of brush thickness on the adsorption of the second, oppositely charged PEL layer for strong/strong and strong/weak PEL systems was investigated. In a set of experiments the brush thickness was varied from 5 nm to roughly 200 nm by adjusting polymerization time,

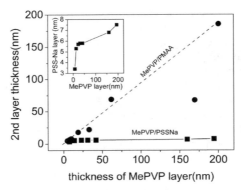

Fig. 28 Layer thickness increase due to formation of a PEL–PEL complex during exposure of a MePVP brush to a PSSNa solution or a PMAA solution. *Lines* are a guide to the eye

which controls the conversion of the initiator and, accordingly, the grafting density of the polymer monolayer. Figure 28 shows how the amount of polyanions adsorbed depends on the polycation (MePVP) brush thickness. It is evident that the adsorption of PMAA on the MePVP brush strongly depends on the initial brush thickness whereas the adsorption of PSSNa is almost independent of the thickness of the substrate brush. While in the latter case the brush thickness was varied by a factor of about 20, the concurrent change of the thickness of the absorbed PSSNa layer was only a factor of about 1.5. In contrast to this, for adsorption of the weak polyelectrolyte (PMAA) the thickness of the adsorbed PMAA layer is more or less identical to the thickness of the brush on which it was adsorbed, and an increase in the layer thickness by a factor of roughly 20 was observed. This difference in the overall layer stoichiometry can be understood if the swelling of the layers in contact with the oppositely charged polyelectrolyte is studied with multiple-angle ellipsometry. Whereas the strong-weak PEL system remains highly swollen throughout the absorption experiment, the brush in the strong/strong system collapses immediately upon addition of the oppositely charged PEL.

Apart from the formation of ultrathin surface-attached PEL–PEL complexes it is very interesting whether the PEL brushes can be also used for the formation of PEL multilayer assemblies. The so-called layer-by-layer (LBL) technique is a simple and powerful method to form well-defined multilayered structures [80]. For the formation of such multilayer assemblies the brushes are dipped alternately into polyelectrolyte solutions, one consisting of a positively charged polyelectrolyte, the other of a negatively charged polyelectrolyte. It is usually assumed that in this LBL deposition process the driving force for each monolayer formation is charge overcompensation [81, 82]. The stability of the multilayered system formed by LBL process in different environments is one of the limitations of this process. Since the attachment of the first layer depends solely on the interaction of

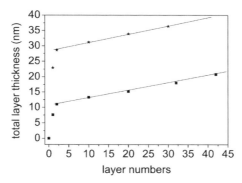

Fig. 29 Film thickness as a function of the layer numbers for MePVP/PSSa multilayers using 7.6 nm (*squares*) and 22.9 nm (*stars*) MePVP covalently attached monolayers as the first layer. The *solid lines* show a linear fit of the dependence of the film thickness on the number of deposited layers

the polymers with surface charges, the whole multilayer assembly can be desorbed either by changing the sign of the surface charge of the substrate or by addition of competing low-molecular-weight electrolytes which can displace the polymer molecules in the first monolayer [82, 83]. Here having covalently anchored polymers could be advantageous.

Another general problem of the layer-by-layer method is the low thickness of each single deposited layer, which is on the order of 0.5 nm [80]. Such a small increase of the film thickness per deposited layer is rather inconvenient if a thicker PEL multilayer assembly is desired, as in this case many layers have to be deposited.

In preliminary experiments on the formation of polyelectrolyte multilayers deposition of a MePVP/PSSNa multilayer system and a MePVP/PMAA multilayer system were carried out using systems such as those described above. The results of this study are shown in Figure 29 and Figure 30. The prior shows the results of experiments in which two strong polyelectrolytes were used (Fig. 29). It can be easily be seen that the thickness increase due to absorption of the second layer is larger than that of the following layers deposited on top of it, but starting from the third layer the thickness increases only by about 0.5 nm per deposition cycle (i.e. deposition of two monolayers) and a linear relationship between the layer thickness and the number of deposited layers is observed. In this system the increase in layer thickness per deposition cycle is independent of the properties of the brush.

For the MePVP/PMAA system also (Fig. 30) it is clearly visible that the film thickness increases linearly with the number of dipping cycles. However, it is also evident that the thickness of each layer in the multilayer assembly strongly depends on the thickness of the initial brush layer. The increase of layer thickness per deposition cycle is more or less identical to that of the thickness of the initial brush monolayer and even when a very thick multi-

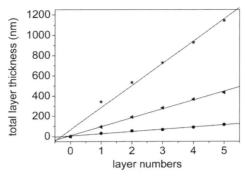

Fig. 30 Film thickness as a function of the layer numbers for MePVP/PMAA multilayers using 31 nm (*circles*), 90 nm (*triangles*) and 342 nm (*stars*) MePVP brushes as the first layer. The *solid lines* represent a linear fit of the dependence of the film thickness on the number of deposited layers

layer is deposited, the outermost layer resembles closely the innermost, the brush layer, and a very strong templating effect is observed.

Although the overall architecture of the two systems is very similar (surface-attached PEL brush with electrostatically attached monolayers of alternating charge sign), the film formation behavior of the two systems is very different, as shown schematically in Fig. 31. In the depiction shown in the figure only the heigth increase per layer is schematically shown. The figure is not ment as to imply any detail concerning the internal structure such as the existence or absence of phase boundaries and on film roughnesses. Although all the details of this difference are not yet understood, it is evident that the basic difference between the systems is the water solubility of the PEL–PEL complex formed. This difference is directly evident if solutions of the two polyelectrolytes are mixed [84]. While in the first (weak/weak) case the complex remains soluble if solutions containing equimolar amounts of

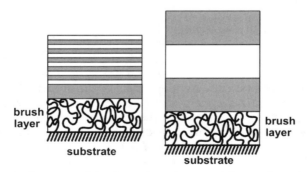

Fig. 31 Schematic depiction of the formation of PEL multilayers through PEL brushes: (a) strong/strong system; (b) strong/weak system; no implication is made about phase boundaries between the different layers and interface roughness

polyanion and polycation are mixed, in the latter (strong/strong) case immediate precipitation occurs due to the hydrophobicity of the neutral complex formed.

Polyelectrolyte brushes can be used as substrates for the absorption of polyelectrolytes from solution to form PEL–PEL complexes attached to solid surfaces and for the subsequent build-up of weak polyelectrolyte multilayers by a layer-by-layer deposition process. Because the generated polyelectrolyte monolayers directly in contact with the surface are attached to it by chemical bonds, the PEL complexes and the PEL multilayer could be potentially very stable. Another advantage of this method is that it allows control of the surface properties of the substrate in great detail.

The formation of PEL complexes and multilayers at solid surfaces via polymer brushes consisting of weak and strong PELs is quite different from the case where two strong polyelectrolytes are used. Interestingly, in the strong-weak PEL system, each adsorbed layer has the same thickness, which closely resembles the thickness of the innermost, namely the brush layer, whereas in the strong/strong PEL system the formation of PEL multilayers resembles more the traditional LBL technique in so far as that the increase in layer thickness per deposition cycle is well below 1 nm. In the strong/weak case, however, easily more than 100 nm of polyelectrolyte can be deposited per dipping cycle, which allows the simple generation of very thick PEL multilayer assemblies.

4
Mixed Polyelectrolyte Brushes: Adaptive and Responsive Polymer Surfaces

4.1
Introductory Remarks

A novel strategy for the fabrication of responsive functional polymer films is based on grafting of several different functional polymers on to a solid substrate via functional end groups, resulting in mixed polymer brushes. Grafting of two different polymers randomly on to a surface prevents macrophase separation while numerous morphologies and the possibility of switching between them as a result of external stimuli offer opportunities to tune surface properties and structure in a wide range. Even more promising properties can be attributed to the mixed brushes if one or more polymers in the mixed brush carry electrical charges.

4.2
Theory of Mixed Brushes

Mixed brushes prepared from polymers of similar molecular weight but different chemical composition have recently been intensively investigated theoretically [85–89] and experimentally [89–92]. It was shown that the mixed brush morphology is effected by the interplay between lateral and phase segregation governed by solvent quality. The lateral segregation of mixed brushes is dominant in nonselective solvent and results in a ripple morphology, while in selective solvent the unfavored polymer forms clusters embedded in the continuous phase (dimple morphology) of the second polymer with its enhanced concentration on the top of the layer. This mechanism introduces the adaptive and switching properties of the thin film which can change morphology, surface energetic state, and functionality upon exposure to controlled environments.

Mixed brushes fabricated by use of oppositely charged polyelectrolytes (PEL) grafted on the same planar substrate exhibit even much more complicated behavior [93]. In contrast with equally charged homopolymer brushes, for which intrachain and interchain Coulombic repulsion leads to a stretching of the chains, oppositely charged mixed brushes have other means of reducing the electrostatic repulsion, depending on charge ratio (degree of compensation of the total charge of homopolymer A, consisting of N_A segments, by the total opposite charge of homopolymer B, consisting of N_B segments). When $N_A > N_B$ polymer A chains are coiled due to the electrostatic attraction between A and B, whereas when $N_A = N_B$ both chains form a compact brush. Addition of salt leads to screening of this attraction and the brush expands. At a low charge ratio further increase of salt concentration causes a decrease of brush thickness due to screening of the repulsion between equally charged segments, thus exhibiting a maximum. Consequently, the larger variety of combinations of electrostatic and short-range interactions leads to a much more complex response.

4.3
Synthesis of Mixed Polyelectrolyte Brushes

The strategy for synthesis of mixed PEL brushes is based on a combination of the concepts used for synthesis of homopolymer PEL brushes, when two main "grafting to" [94–99] and "grafting from" [63, 64, 66, 100] approaches

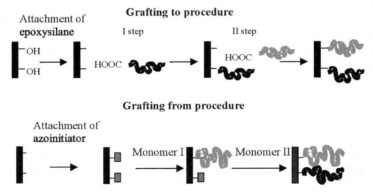

Fig. 32 Schematic presentation of "grafting to" and "grafting from" approaches for synthesis of binary polymer brushes

were developed by a number or research groups, and of the strategy developed for synthesis of mixed polymer brushes [80–92]. Usually neutral homopolymer brushes are synthesized by "grafting to" or "grafting from" method and the second homopolymer is then grafted in the next step (Fig. 32). Finally, the mixed polyelectrolyte brush is prepared from the mixed neutral brush with an appropriate polymer analogous reaction. Below we present several examples of different strategies for grafting of mixed PEL brushes.

4.3.1
"Grafting To" Method. Step by Step Grafting of Two End-Functionalized Homopolymers

Mixed Polystyrene (PS)—poly(2-vinylpyridine) (P2VP) brushes were synthesized by grafting carboxyl terminated PS in the first step and carboxyl terminated P2VP in the second [92]. The mixed brush was prepared on the surface of an Si wafer functionalized with (3-glycidoxypropyl)trimethoxysilane (GPS). In both cases the polymers were spin-cast on the substrate and grafted from the melt at a temperature above the T_g of the polymers (120 °C–150 °C) via reaction of surface epoxy and hydroxyl groups with end carboxyl groups of the homopolymers. In the next step the mixed PEL brush was obtained by protonation of P2VP in acidic aqueous solution. Since the grafted amount of the end-functional polymers is kinetically limited, the composition of the brush was regulated by the conditions of grafting in the first step (temperature and time) to graft some limited amount of the first homopolymer and to reserve space on the surface for grafting of the second polymer. In most cases the two different polymers are incompatible and, generally, the best results for the grafting procedure on the Si wafer were obtained if the second polymer shows larger affinity to the substrate than the first one. For the inverse case the preferential adsorption of the first homopolymer protects the surface functional groups from grafting of the second polymer.

Polyacrylic acid (PAA)—P2VP mixed brushes were prepared by a similar synthetic procedure, by grafting of carboxyl-terminated poly(*tert*-butyl acrylate) (PtBuA) and P2VP. Afterwards, PtBuA was hydrolyzed in the presence of *p*-toluene sulfonic acid. The same strategy was employed to graft mixed PEL brushes on polymer surfaces. In this case plasma treatment was used to functionalize surface of polymer substrates. We introduced amino groups on the surface of PA-6 and PTFE by treatment of the polymer samples with NH_3 plasma. Then the carboxyl terminated homopolymers were grafted step by step from the melt to the solid substrate via amide bonds.

Mixed brushes of different chemical compositions and ratios between polymers can be prepared with this procedure (Table 1). However, the total grafted amount is usually limited to 10 mg m^{-2}.

Table 1 Characteristics of mixed PE brushes

Method and composition	Grafting to PS-P2VP			Grafting to PtBuA-P2VP			Grafting from PS-P2VP		
	PS-P2VP	PS	P2VP	PtBuA-P2VP	PtBuA	P2VP	PS-P2VP	PS	P2VP
Grafted amount A (mg m^{-2})	6.8	3.6	3.2	7.3	4.0	3.3	27.3	14.1	13.2
Grafting distance (nm)	3.2	4.6	4.3	3.0	4.2	4.5	3.3	4.9	4.5
Grafting density (nm^{-2})	0.11	0.047	0.054	0.11	0.06	0.05	0.09	0.04	0.05
M_n (kg mol^{-1})		45.9	39.2		42.0	39.2		197.7	162.5
M_w (kg mol^{-1})		48.4	41.5		47.0	41.5		361.5	284.8

4.3.2
"Grafting from" Method

This method can be successfully used to prepare high grafting density mixed PEL brushes with a similar strategy [90, 91]. The surface of an Si wafer is functionalized with GPS and treated with diaminopropane. The acid chloride of 4,4′-azobis(4-cyanopentanoic acid) (Cl-ABCPA) was chemically grafted to hydroxyl and amino groups on the Si wafer surface. Step by step two different monomers, styrene and 2-vinylpyridine, were polymerized on the surface. The ratio between the two grafted polymers is regulated by the conditions of polymerization (temperature and time) in each step of grafting. In the second step polymerization is initiated by the residual fraction of the chemisorbed azo-initiator left after the first step. The molecular weight of the grafted polymers can be regulated by adjusting monomer concentration, temperature, and concentration of initiator added to the bulk solution.

The same protocol was used for the "grafting from" procedure to graft mixed brushes on to polymeric substrate. Plasma treatment was used to introduce hydroxyl (oxygen plasma) and amino (NH$_3$ plasma) functional groups on to the polymeric surfaces and then Cl-ABCPA and two different homopolymers were grafted by polymerization from the functionalized surface.

4.4
Characterization of Mixed PEL Brushes

Thickness of the brush and surface roughness of the layers was evaluated by X-ray reflectivity which we performed with the dried product after each grafting step. Results of all methods agreed very well providing evidence for the reliability of the methods and models for fitting [90, 92].

4.4.1
Phase Segregation and Concentration Profiles

Mixed brushes segregate laterally and perpendicular to the plane of the substrate. That introduces the difficult task of revealing the three-dimensional concentration distribution of each component in the brush in various media. It was impossible to solve this problem directly and we subdivided the problem into two tasks, i.e. to reveal first the concentration profile in the Z-direction and then study the lateral segregation in the layer. Even this divided task is quite difficult because of the complicated morphology. For simplicity we perform quenching of the brush layers after treatment with particular solvents, assuming that the lateral segregation is not changed during the rapid evaporation of the solvent and the brush undergoes mainly collapse in Z-direction, while in X–Y directions the morphology is only slightly changed. Good agreement between the results obtained from different methods proves the validity of this approach.

The lateral chemical and topographical resolution of the morphology was investigated with atomic-force microscopy (AFM) and X-ray photoemission electron microscopy (XPEEM).

XPEEM is a spectromicroscopy technique utilizing near edge X-ray absorption fine structure (NEXAFS) contrast. XPEEM probes the surface chemistry by detection of photoelectrons emitted from the surface. XPEEM was performed at the PEEM2 microscope at the bending magnet beamline 7.3.1.1 of the advanced light source synchrotron at the Ernest Orlando Lawrence Berkeley National Laboratory [89]. We estimated that the lateral resolution provided by the microscope for the polymer samples was better than 70 nm. *Surface charge* averaged over the sample was measured with a Centec PAAR Physica Apparatus in a 0.001 mol L^{-1} KCl solution.

4.5
Switching/Adaptive/Responsive Properties

Large diversity of functional surfaces can be fabricated with approach of mixed PEL brushes. Here we consider two most important examples (Fig. 33):

1. PEL–non PEL mixed brush; and
2. mixed brush from two different oppositely charged PEL.

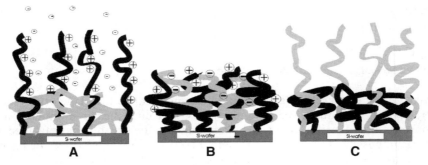

Fig. 33 Illustration of a polymer brush behavior under different conditions: (a) PE-non-PE and PE1-PE2 brush in acidic media, (b) PE1-PE2 near IEP and (c) PE-non-PE exposed to a selective organic solvent for the non-polar component

4.5.1
PeI–Non PeI Mixed Brush

This brush can be considered as a limiting case of the mixed PEL brush of the homopolymers A and B at the charge ratio equal to 0. Responsive properties are determined from the combination of the behavior of the uncharged mixed brush and the additional behavior originating from the electrostatic interchain and intrachain repulsion of chains of the charged homopolymer A. Chains of the polymer B display coiled conformation and are located near the substrate surface, they are in bad solvent conditions. The remaining chains of the charged polymer A are stretched. The main result of such a combination lies in an extension of the range of switching—charges make the surface more hydrophilic and introduce specific electrostatic interactions important for adsorption of colloidal particles, proteins and cells. The responsive properties of a weak PEL–non PEL brush can be regulated by change of pH. Below we present several examples.

The Isoelectric point (IEP) of the mixed PS-P2VP brush, grafted on to GPS-modified Si wafer, was determined to be at pH 5.9. That is between IEPs of the homopolymer brushes of PS (pH 4.3) and P2VP (6.7). For a PtBuA–P2VP mixed brush the IEP was at pH 5.7 (IEP for PtBuA at pH 4.6). Both brushes undergo switching of morphology from dimple to ripple regime (Fig. 34) and similarly switch the surface energetic state (wettability) upon exposure to different organic solvents and aqueous medium of different pH (Table 2). Upon exposure to toluene the brushes are hydrophobic. The wettability of the brushes increases with decrease of the pH. At low pH values the top of the film is preferentially occupied by hydrophilic protonated P2VP chains. As reference we added the wetting properties of the two homopolymer brushes PS (non PE) and P2VP (weak PE). It is clearly seen that the electrostatic interactions extend switching of wettability of the brushes to broader range from highly hydrophilic (contact angle 22°) to hydrophobic (contact angle 90°).

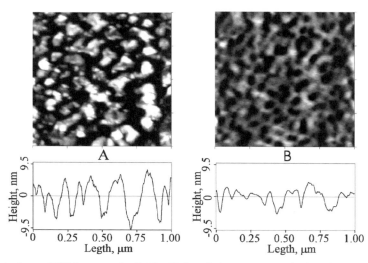

Fig. 34 1×1 μm AFM images of a PS-P2VP brush (1:1, 24.7 nm thick) after exposure to different aqueous media: (**a**) pH=1.6, (**b**) pH=9.3

Table 2 Water contact angles $(\theta \pm 3)^0$ of binary (1:1) PS-P2VP and PtBuA-P2VP brushes

Sample	PS–P2VP	PS–P2VP	PPtBuA-P2VP	PS	P2VP
A (mg m^{-2})	27.2	6.9	7.3	6.8	7.2
Water pH					
1.51	22	41	–	–	19
2.9	58	55	57	87	24
5	–	–	69	–	–
7	67	71	68	86	62
9.3	73	72	73	88	63
Methanol	61	64	62	88	66
Toluene	90	88	82	90	70

4.5.2
Oppositely Charged Mixed Brushes

In this type of mixed brush one may expect also quite large diversity of combinations. Here we present one example of the mixed brush fabricated from oppositely charged weak PEs—the PAA–P2VP mixed brush. The IEP point of the brush (pH 5.6) is in between those of the corresponding homopolymer brushes (PAA pH 3.2; P2VP pH 6.7) indicating the area when the brush is neutral and opposite charges compensate each other in the brush.

The mixed PEL brush is sensitive to pH in a wide range of acidic–basic properties of solution. That we illustrate with the dependence of the brush thickness vs. pH. This relationship is presented by a symmetric U-shaped

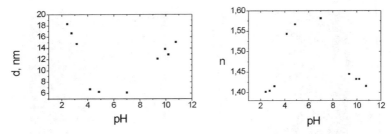

Fig. 35 Influence of pH on the thickness, d, and refractive index, n, of the swollen PAA-P2VP brush. The thickness of the dry state was measured to be 6.4 nm

plot (Fig. 35). The effective refractive index of the swollen brush shows an inverse relationship.

4.6
Concluding Remarks on Mixed Brushes

The use of mixed polyelectrolyte brushes is proposed as a novel approach for engineering of responsive surfaces. Grafting of different functional polymers on to the same surface prevents macrophase separation and offers exclusive possibilities to use the interplay between intermolecular and intramolecular interactions to tune properties of the thin polymer films–surface composition, surface energetic state, surface charge, thickness, wettability, adhesion, etc., can be changed in a wide range. The experimental results confirm that the concept of mixed PEL brushes opens a very promising route for the development of new functional materials for application in different areas. One intriguing aspect is, for instance, that those mixed brushes can react both on aqueous media as well as on organic solvents, while they show switching behavior on pH and salt. Together with the wide variability in different polymer brushes this opens a huge playground for different properties and their design.

5
Cylindrical Polyelectrolyte Brushes

5.1
Introduction

Typical linear polyelectrolyte chains exhibit a charge density of one ionic charge per 0.25 nm contour distance or less. According to Oosawa [101, 102] for charge densities higher than one charge per Bjerrum length, l_B, given as:

$$l_B = e^2 / (4\pi\varepsilon_o\varepsilon K_B T) \tag{7}$$

where e is the electronic charge, ε_0 and ε the dielectric permittivity in vacuo and in solution, respectively, and $K_B T$ the thermal energy, the counter-ions

Table 3 Comparison of different chemical pathways to prepare cylindrical brushes

	Grafting on to	Grafting from	Macromonomer polym.
Main chain polydispersity	+	+	−
Side chain polydispersity	+	0	+
Control of grafting density	−	0	+

"condense" on to the polyion chain (known as "Mannig condensation"), thus reducing the effective charge density to a maximum of one charge per Bjerrum length. Whereas the existence of counter-ion condensation is no longer questioned, the condensation threshold and the chemical nature of the "condensed ion pair" are still controversially discussed. In particular, the flexibility and hydrophobicity of the polyelectrolyte chain, the chemical nature of the counter-ions, the solvent quality, and concentration effects may well influence the "Manning condensation" [103–105].

In polyelectrolyte networks and in planar polyelectrolyte brushes the degree of swelling may be caused by electrostatic repulsion of the charged chains or chain segments for small charge densities whereas for high charge densities condensation of counter-ions causes strong osmotic swelling due to the Donnan equilibrium. In the latter case the electrostatic interactions are screened by the presence of the condensed counter-ions. In this chapter we review some properties and recent synthetic progress in the preparation of cylindrical polyelectrolyte brushes as novel hybrid structures between linear polyions and planar polyelectrolyte brushes.

Since the pioneering work of Tsukahara on the homopolymerization of methacryloyl end-functionalized polystyrene macromonomers to polymacromonomers several chemically different polymacromonomers were successfully synthesized [106–118]. Also different routes to cylindrical brush polymers by "grafting from" [119–121] and "grafting on to" [122, 123] techniques were reported, each of which exhibits certain advantages and disadvantages, as outlined in Table 3.

In the following text we focus on cylindrical brushes prepared by polymerization of macromonomers, in particular on two systems. First, cylindrical brushes consisting of poly-2-vinylpyridine side chains and a polymethacrylic main chain are discussed; these were converted into cationic polyelectrolyte brushes by quaternization of the pyridine units by alkyl halides. The second system comprises sulfonated anionic cylindrical brushes, i.e. polymacromonomers with polystyrene side and main chains and polymacromonomers with polymethacrylic main and polymethacrylic acid side chains.

5.2
Cylindrical Brushes with PVP Side Chains

5.2.1
Neutral Brushes

The macromonomers were prepared by anionic polymerization of 2-vinylpyridine followed by reaction with ethylene oxide and methacrylic acid chloride [111] as shown in Scheme 1. MALDI–TOF mass spectroscopy was utilized in order to determine the absolute molar mass and the degree of end-functionalization as given in Table 4. The sample code MM-PVPXY comprises the polymerizable unit (MM=methacrylate), the side chain (PVP=polyvinylpyridine) and the side chain degree of polymerization XY.

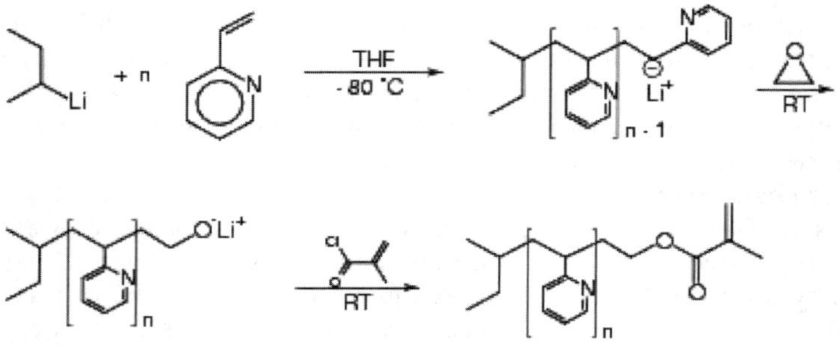

Scheme 1

The radical polymerization was performed at high macromonomer concentration in benzene as described elsewhere [111] in order to yield high molar mass polymers. The resulting polymacromonomers were characterized by AFM and by static and dynamic light scattering. One typical AFM-picture is given in Fig. 36 which demonstrates the polymacromonomers exhibit the conformation of wormlike cylindrical brushes with a considerable directional persistence. The extensional force is caused by the strong steric repulsion between the densely grafted side chains. Light scattering in typical

Table 4 MALDI–TOF characterization of PVP macromonomers

Sample	M_n	M_w	M_w/M_n	$f(\%)$ [a]
MM-PVP 21	2352	2521	1.07	n.d. [b]
MM-PVP 25	2846	3188	1.12	n.d.
MM-PVP-26	2936	3081	1.05	85
MM-PVP 36	3952	4191	1.06	n.d.
MM-PVP 47	5078	5281	1.04	75

[a] degree of endfunctionalization
[b] not determined

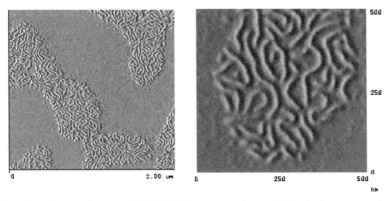

Fig. 36 AFM picture of sample PM-PVP47A spin cast from THF solution on to mica

solvents like THF or methanol could not be utilized for molar mass characterization because anomalous Zimm-plots were observed, as shown in Fig. 37. Since the concentrations investigated are well below the overlap concentration, c^*, the interaction peak is believed to originate from long-range electrostatic repulsion caused by ionic impurities in the polymer. Measurements in DMF with 1 g L^{-1} LiBr added yielded normal scattering curves (Fig. 38) thus corroborating the presence of ionic charges even in the non-quaternized polymer. The characterization results are summarized in Table 5. As compared to the macromonomer the sample code of the polymacromonomer is changed to PM-PVPXY.

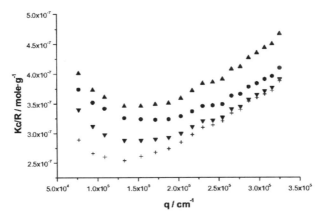

Fig. 37 Anomalous scattering envelopes caused by intermolecular scattering for different concentrations of sample PM-PVP47A in THF: *circles*, 1.150 g L^{-1}; *triangles*, 0.61 g L^{-1}; *inverted triangles*, 0.49 g L^{-1}; *plus signs*, 0.115 g L^{-1}

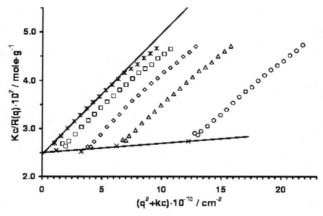

Fig. 38 Zimm plot of sample of PM-PVP47A in DMF with 1 g L^{-1} LiBr added: *circles*, 0.245 g L^{-1}; *triangles*, 0.125 g L^{-1}; *diamonds*, 0.066 g L^{-1}; *squares*, 0.024 g L^{-1}

Table 5 Light-scattering characterization of PVP-polymacromonomers

Sample	Macromonomer	$M_w \times 10^{-6}$	R_g (nm)	R_h (nm)	M_w/M_n[a]
PM-PVP 21	PM-PVP 21	2.7[b]	70[b]	34[b]	3.63
PM-PVP 25	PM-PVP 25	0.29	19.4	10.7	3.07
PM-PVP26	PM-PVP 26	5.4	68.4	45.2	5.3
PM-PVP36	PM-PVP 36	1.8	27.6	20.5	3.14
PM-PVP47A	PM-PVP 47	4.2	54.9	36.3	5.22
PM-PVP47B	PM-PVP 47	2.94	46.1	29.4	5.2
PM-PVP47C	PM-PVP 47	2.3	39.6	29.6	3.9

[a] Determined by GPC with polystyrene calibration
[b] Approximate values, only, due to strongly curved Zimm-plots and a negative virial coefficient

5.2.2
Quaternized Brushes

Cylindrical polyelectrolyte brushes were prepared by quaternization of the respective PVP-brushes with alkyl bromides. For the quaternized samples the sample code is supplemented by the alkylation agent (Me=methyl, Et=ethyl, Bz=benzyl, and Do=dodecyl) followed by the degree of quaternization, Q. The degree of quaternization was determined by IR and is listed in Table 6 along with the light-scattering results. Typical Zimm-plots for sample PM-PVP47C-Et75 in 10^{-3} mol L^{-1} NaBr and 10^{-2} mol L^{-1} NaBr are shown in Fig. 39. Light-scattering investigations in water as a function of added salt did reveal a small decrease in the molecular dimensions, which hardly exceeded the experimental error of ±10% for M_w and R_g and ±5% for R_h. This increase is much less than observed for linear flexible polyvinylpyridinium

Table 6 Light-scattering and IR characterization of cylindrical polyelectrolyte brushes prepared by quaternization of PVP brushes

Sample	Akylation agent	Q^a	R_g/R_h (nm)			
			10^{-1}m	10^{-2}m	10^{-3}m	10^{-4}m
PM-PVP21-Bz80	Benzyl bromide	80	–	–	70/39.4	77/42.5
PM-PVP25-Me80	Methyl bromide	90	21.5/11.5	21.5/11.5	–	–
PM-PVP26-Et90	Ethyl bromide	90	–	92/53	105/56	100/61
PM-PVP36-Et70	Ethyl bromide	70	31.4/21.4	35.9/24.2	38.8/26.3	–
PM-PVP36-Do50	Dodecyl bromide	50	–	–	–	–
PM-PVP47C-Et75	Ethyl bromide	75	–	42/30	48/32	52/39

[a] Degree of quaternization

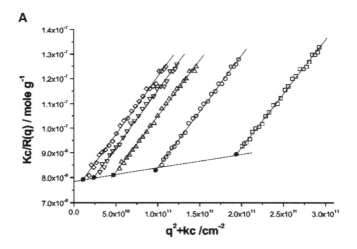

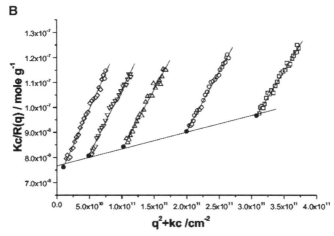

Fig. 39 Zimm plot of sample PM-PVP47C-Et75 in 10^{-2} mol L^{-1} NaBr (**a**) and 10^{-3} mol L^{-1} NaBr (**b**) solution

chains of similar chemical composition, because the excluded volume expansion is drastically reduced for the cylindrical brushes which exhibit a wormlike conformation. However, due to electrostatic repulsion and/or osmotic swelling a significant increase of the dimensions was to be expected but was not found experimentally. The cylindrical polyelectrolyte brushes could be stiffened at constant contour length or the contour length may increase with or without a change in chain stiffness. The latter scenario was found for neutral brushes for which the contour length increased significantly with increasing solvent quality [124]. The scatter of the data and the small changes in the dimensions do not allow the possibilities discussed above to be distinguished.

Depending on the hydrophobicity of the quaternization agent the polyelectrolyte brushes start to aggregate at higher salt concentrations and eventually precipitate. This aggregation has been observed "in situ" by AFM for sample PM-PVP21-Bz80 [125].

5.2.2.1
The Cross-Sectional Dimensions

In order to determine the effect of the ionic charges on the cross-sectional dimensions, i.e. on the extension of the side chains, X-ray scattering experiments were performed. In the high q -regime the form factor of a rod-like structure decays as:

$$q \cdot P(q) = \pi/L \cdot \exp\left(-\frac{1}{2}q^2 R_{g,c}^2\right) \tag{8}$$

where L is the length of the rod and $R_{g,c}$ the cross-sectional radius of gyration. In Fig. 40 the cross-sectional radius is plotted vs added salt concentra-

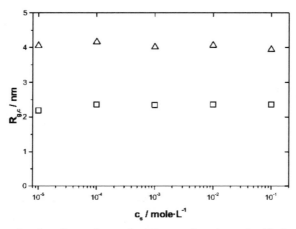

Fig. 40 Cross-sectional radius of gyration $R_{g,c}$ as function of added salt for sample PM-PVP25-Me90 (*squares*, NaBr) and for sample F3-PS-PS41-S66 (*triangles*, NaCl)

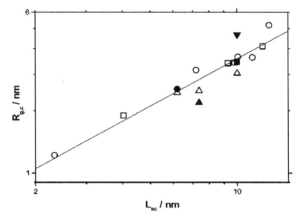

Fig. 41 Cross-sectional radius of gyration $R_{g,c}$ as function of the side chain length L_{sc}, for neutral and ionic cylindrical brushes: *open circles*, PM-PS samples in toluene, SANS; *open squares*, PM-PS samples in toluene, SAXS; *open triangles*, PM-PVP samples in MEK, SAXS; *filled circles*, PM-PVP21-Bz80; *filled triangles*, PM-PVP25-Me90; *filled squares*, PM-PVP36-Et70; *filled inverted triangles*, PM-PVP36-D$_O$50

tion for sample PM-PVP25-Me90 and for the polystyrenesulfonate cylindrical brush as discussed above. Within experimental error the cross-sectional dimension does not change with added salt from "salt free" condition up to 10^{-2} mol L^{-1} salt. These results document that electrostatic repulsion and/or osmotic swelling does not significantly affect the side-chain expansion. Also, the ionic side chains are apparently not more expanded than the neutral counterparts, as shown in Fig. 41. Those results seem to confirm the invariance of the dimensions upon ionization. The solid line in Fig. 41 represents a least square fit to all data yielding a slope of 0.75. This value lies at the upper limit of what is theoretically expected.

5.2.2.2
Effective Charge Density

In order to determine the effective charge density of the polyelectrolyte brushes conductivity measurements were performed under argon gas atmosphere on salt-free aqueous solutions. For evaluation of the effective charge density the mobility of the polyion and of the counter-ion have to be known according to:

$$f = \frac{\sigma}{c_p^m \left(\mu_p + \mu_c \right)} \tag{9}$$

where σ is the bare conductivity due to polyion and counter-ions, c_p^m the concentration of monomer units (in "monomole L^{-1}"), $e=1.602\times10^{-19}$ C the elemental charge and μ_p and μ_c are the respective mobility of polyions and

Table 7 Fraction of free counterions, f_c (normalized to the number of chemically quaternized monomers), the effective charge density, f (normalized to the total number of monomers), the effective charge density per main chain monomer f_{mc}, the cross sectional radius of gyration $R_{g,c}$, the mean concentration of counterions c_c and the mean inverse Debye screening length χ_b^{-1} within the volume of a cylindrical brush molecule due to condensed counterions

Sample	f_c (%)	f (%)	f_{mc} (%)	$R_{g,c}$ (nm)	χ_b^{-1} (nm)	c_c (mol L^{-1})
PM-PVP 21-Bz80	3.3	2.6	55	2.55	0.24	6.65
PM-PVP 25-Me90	14.9	13.4	348	2.36	0.21	8.55
PM-PVP 36-Et70	7.9	5.5	198	3.45	0.25	6.04
PM-PVP 36-Do50	4.4	2.2	79	4.66	0.33	3.43
lim PVP-Et70	28.0	19.6	19.6	–	–	–

counter-ions. Practical problems are the subtraction of the conductivity of pure water, which is quite sensitive to small amounts of ionic impurities, and the formation of HBr upon dissolution of the polyions because the pyridinium groups represent a weak base. Thus, the results rely on a precise determination of the pH value since protons exhibit the highest electrophoretic mobility ($\mu_{H^+} = 36.23 \cdot 10^{-4}$ and $\mu_{Br^-} = 8.09 \cdot 10^{-4} cm^2 s^{-1} V^{-1}$). The net conductivity of the polyions plus counter-ions is given by:

$$\sigma = \sigma_{sol} - \sigma_w - \sigma_{HBr} \tag{10}$$

where σ_{sol}, σ_w, and σ_{HBr} are the respective conductivities from the solution, from the solvent (water), and from HBr. According to Volk [126] the polyvinylpyridinium conductivities at low salt concentrations lie in the range 5×10^{-4} cm^2 s^{-1} V^{-1} which was utilized for the calculation of f. In Table 7 the effective charge density f is given for cylindrical polyelectrolyte brushes with different hydrophobicity and for linear polyions for comparison. The effective charge density for the linear polyions compare reasonably well to the values obtained by Beer et al. [103] for similar samples. It is clearly seen that the effective charge density of the polyelectrolyte brushes is generally very small.

However, it is to be noted that the mobility of the polyelectrolyte brushes is not known, and is also assumed to be $\mu_p \cong 5 \times 10^{-4}$ cm^2 s^{-1} V^{-1}, which may not be justified due to the different chain architecture. Since $\mu_p \approx Z/R_h$ (where Z is the number of ionic charges and R_h the hydrodynamic radius of the polyion) we may estimate the mobility of the brushes to be higher than that of a flexible coil, because the reduced number of charges is overcompensated by the much more compact chain architecture. Thus, the effective charge densities given in Table 7 most probably represent an upper limit. Moreover, the value for the polyion mobility does not influence the result for the effective charge density significantly as long as it is smaller than the conductivity of Br$^-$. Even for the very unlikely case that μ_p is as small as 10^{-4} cm^2 s^{-1} V^{-1} the resulting charge density increases by 50% only. Qualitatively, the following trends become clear. The charge density of a cylindrical

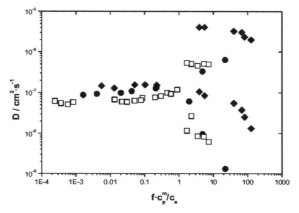

Fig. 42 Fast and slow diffusion coefficients as function of the reduced quantity $fc_p{}^m/c_s$: *squares*, PM-PVP21-Bz80; *diamonds*, PM-PVP25-Me90; *circles*, PM-PVP36-Et70

polyelectrolyte brush is much smaller than that of the corresponding flexible linear polyelectrolyte chain. With increasing hydrophobicity of the polyelectrolyte brush the effective charge density decreases. This qualitatively explains why static light-scattering measurements reveal neither intermolecular structure peaks in dilute solution nor the dynamic coupling effect, along with the occurrence of the yet unexplained slow mode, unless solutions with extremely high polyion concentrations are measured [104]. The latter effect is reported to occur if the added salt concentration decreases below the concentration of ionic charges within the polyion $c^m{}_{p,eff}=fc^m{}_p$ in monomole L^{-1}, i.e. $c^m{}_{p,eff}/c_s<1$. Accordingly, this transition is expected to be shifted to much higher polyion concentration if the charge density is very low [127, 128]. In Fig. 42 dynamic light-scattering measurements are shown as a function of $fc^m{}_p/c_s$ for some cylindrical brush polymers, where the effective charge density f was taken from the conductivity results. It is observed that within experimental error the curves exhibit the "ordinary–extraordinary" transition around $fc^m{}_p/c_s=1$. The scatter of the experimental data is quite large because of the high concentrations utilized. Within the concentration regime investigated the dilute–semidilute solution transition should also occur, which makes interpretation of the data slightly ambiguous.

In summary, the light-scattering investigations support the results obtained by conductivity measurements that the effective charge density of cylindrical polyelectrolyte brushes is much smaller than for linear flexible polyions.

5.3
Cylindrical Brushes with Anionic Side Chains

5.3.1
Neutral Brushes

Anionic cylindrical brushes are accessible either by sulfonation of polystyrene brushes or by hydrolysis of *t*-butyl acrylate brushes. Since sulfonation even under mild conditions is likely to hydrolyze ester linkages, styryl end-functionalized polystyrene macromonomers were synthesized by reacting the living styryl anion with *p*-vinylbenzyl bromide. Much care has to be taken in order to quantitatively remove excess *p*-vinylbenzyl bromide. After characterization by MALDI–TOF (Table 8) the macromonomer was polymerized to the corresponding cylindrical brushes. In order to reduce the polydispersity of the polymacromonomer the sample was fractionated by continuous flow fractionation [129] which yielded some narrow fractions with different molar masses, as listed in Table 9. For preparation of cylindrical brushes with carboxylate groups methacryloyl end-functionalized *t*-butyl acrylate macromonomers were synthesized by GTP according to Scheme 2. In contrast to anionic polymerization the end-functionalization is performed by utilizing the functional group of the initiator. The characterization of the macromonomer with MALDI–TOF was not successful despite many attempts utilizing various matrices. Eventually, the molar mass was

Table 8 Characterization of S-PS and MA-PBMA macromonomers

Sample	M_n	M_w	M_w/M_n	f (%) [a]
S-PS41	4452	4827	1.08	>90
MA-PBMA31	3900 [b]	–	1.13 [c]	95

[a] Degree of endfunctionalization
[b] Mn by vapor pressure osmometry
[c] Determined by GPC (PMMA calibration)

Table 9 Light-scattering results of the neutral polymacromonomers including the fractionated samples (F) in toluene

Sample	$M_w \times 10^{-6}$	R_g (nm)	R_h (nm)	M_w/M_n [a]
PS-PS41	2.3	38.6	25.2	3.02
F1-PS-PS41	1.22	18.4	15	1.28
F2-PS-PS41	2.6	34.1	22.9	1.38
F3-PS-PS41	5.4	51.4	33.4	1.49
PMA-PBMA31	2.6 [b]	42 [b]	25.6 [b]	n.d. [c]

[a] Determined by GPC in toluene with fractions of cylindrical brushes as standard
[b] Solvent for LS: MEK
[c] Not determined due to anomalous elution in THF

determined by vapor pressure osmometry and the polydispersity was estimated by GPC with polystyrene calibration. Polymerization of the *t*-butyl
acrylate brushes was performed as discussed before. The characterization
data are summarized in Table 9.

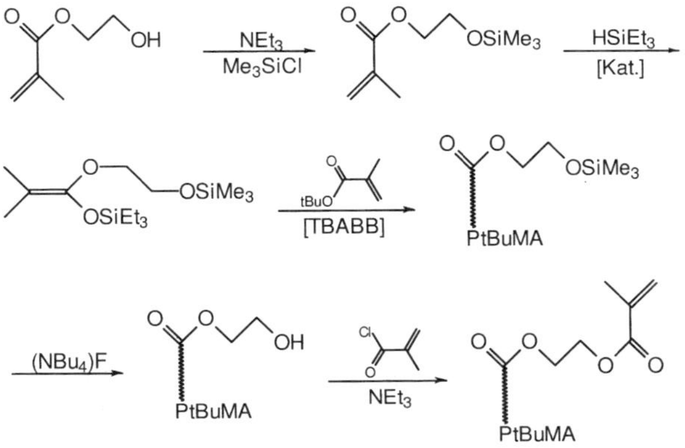

Scheme 2

5.3.2
Anionic Brushes

Although frequently described in the literature, sulfonation of polystyrene
presents a delicate synthetic problem because of the drastic reaction conditions applied. Sulfonation with H_2SO_4/P_2O_5 yielded fully sulfonated brushes.
Light scattering in 10^{-2} mol L^{-1} NaCl solution reveals extremely large molecular dimensions which cannot be explained by a single molecule property
but rather by aggregation. Since this aggregation persists in organic solvents
like formamide it is most probable that side reactions during sulfonation
have led to partial intermolecular cross-linking. Under milder reaction conditions utilizing acetic acid anhydride/phosphoric acid single sulfonated
brushes were obtained with moderate degrees of sulfonation in the order of
50–70% (Table 10). A comparison of the AFM pictures of the neutral precursor brush F3-PS-PS41 (Fig. 43a) and of the sulfonated brush F3-PS-PS41-S68
(Fig. 43b) reveals a more compact conformation of the neutral brushes. This
effect, however, may originate from specific polymer/substrate interactions
and does not necessarily reflect the conformation in solution. It is to be noticed that molar mass and dimensions of the sulfonated brushes are always
significantly larger than those of the respective neutral samples in toluene.
For some samples this is caused by strong aggregation and/or cross linking
which occurs for all sulfonated samples if stored as dry samples in a refrig-
erator for several weeks. Such samples become insoluble, which prohibited
more detailed investigations. Aggregated samples are easily identified be-

Table 10 Light-scattering characterization of cylindrical polyelectrolyte brushes in aqueous 10^{-2} mol L^{-1} NaCl solution

Sample	Q^a	$M_w \times 10^{-6}$ (g mol^{-1})	R_g (nm)	R_h (nm)
PS-PS41-S100	100	12.6	105	71.3
PS-PS41-S64	64	4.6	53	32
F1-PS-PS41-S62	62	2.4	25.8	19.1
F2-PS-PS41-S58	58	6.4	53	34
F3-PS-PS41-S66	66	14.3	140	67
F3-PS-PS41-S68	68	6.85	61.7	36.1
PMA-PBMA31-MAS50	50	n.d. [b]	93.6	52.1
PMA-PBMA31-MAS100	100	n.d. [b]	37.1	22

[a] Determined by titration with NaOH

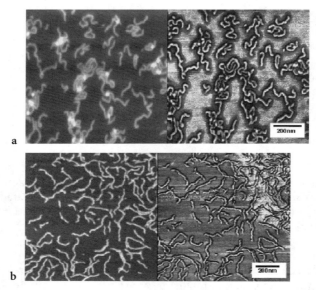

Fig. 43 AFM picture of the neutral precursor brush F3-PS-PS41 (a) and of the sulfonated brush F3-PS-PS41-S68 (b)

cause they exhibit strange scattering envelopes, as shown in Fig. 44 for sample F3-PS-PS41-S66, whereas the non aggregated sample F3-PS-PS41-S68 exhibits normal Zimm Plots (Fig. 45).

Normal Zimm-plots were obtained for samples PS-PS41-S64, F1-PS-PS41-S62, F2-PS-PS41-S58, and F3-PS-PS41-S68 which, however, also exhibit larger molar masses and dimensions as compared to the uncharged precursor samples. In order to trace the origin of this discrepancy the loss of sulfonated polymer during repeated (three times) precipitation by methanol/diethyl ether was investigated for sample PS-PS41-S61. In Fig. 46 the GPC traces of

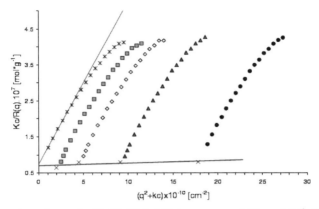

Fig. 44 Zimm-plot of sample F3-PS-PS41-S66 in aqueous 0.01 mol L^{-1} NaCl solution showing partial aggregation: *squares*, 0.013 g L^{-1}; *diamonds*, 0.03 g L^{-1}; *triangles*, 0.061 g L^{-1}; *circles*, 0.119 g L^{-1}

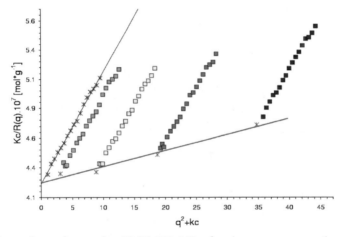

Fig. 45 Zimm-plot of sample F3-PS-P41-S68 showing no aggregation: *squares*, 0.061 g L^{-1}; *circles*, 0.054 g L^{-1}; *triangles*, 0.046 g L^{-1}; *inverted triangles*, 0.3 g L^{-1}; *diamonds*, 0.02 g L^{-1}

precipitated polymer and the fraction isolated from the sol by evaporation of solvent are shown. This clearly shows that a lower molar mass fraction is lost during the purification procedure.

The hydrolysis/elimination reaction of the *t*-butyl methacrylate brushes is not easily driven to full conversion. Whereas linear *t*-butyl methacrylate chains are fully converted to polyacrylic acid in dioxane with catalytic amounts of toluene sulfonic acid the cylindrical brush structures precipitated at about 50% conversion (sample PMA-PBMA31-MAS50). Complete cleavage of the ester groups was achieved, however, by dissolving the partial-

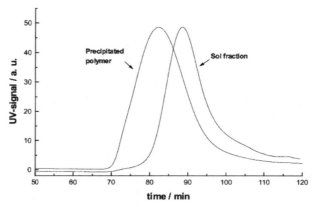

Fig. 46 GPC traces of the precipitated part of sample PS-PS41-S64 and of the sol fraction

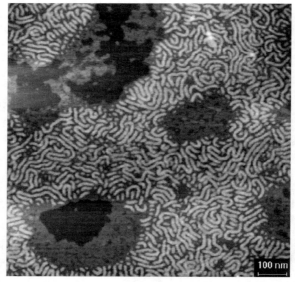

Fig. 47 AFM picture of the partially hydrolyzed sample PMA-PBMA31-MAS50 spin cast from aqueous 0.01 mol L^{-1} NaBr solution onto mica

ly hydrolyzed sample PMA-PBMA31-MAS50 in methanol and subsequent addition of HBr. Both samples were visualized by AFM as shown in Figs. 47 and 48 for PMA-PBMA31-MAS50 and PMA-PBMA31-MAS100, respectively. For the former sample normal conformations of the brushes were observed when spin cast from 0.01 mol L^{-1} NaBr solution on to Mica. The fully hydrolyzed brushes spin cast from pure water at pH≈4–5 showed undulations as sometimes observed for pure PMMA brushes. The origin of such undula-

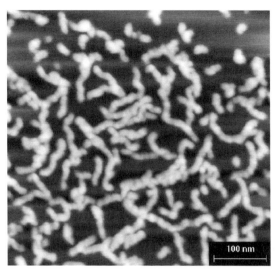

Fig. 48 AFM picture of the fully hydrolyzed sample PMA-PBMA31-MAS100 spin cast from pure water (pH≈4–5) onto mica

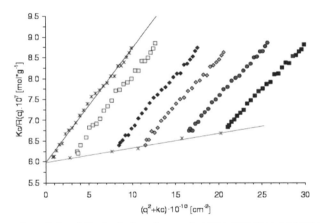

Fig. 49 Zimm-plot of the fully hydrolyzed sample PMA-PBMA31-MAS100 in aqueous 0.01 mol L^{-1} NaBr solution: *squares*, 0.056 g L^{-1}; *diamonds*, 0.153 g L^{-1}; *triangles*, 0.214 g L^{-1}; *circles*, 0.315 g L^{-1}; *plus signs*, 0.405 g L^{-1}

tions is not yet clear and the phenomenon needs further more detailed investigation. Light scattering reveals that the partially hydrolyzed polymers tend to aggregate whereas the fully hydrolyzed sample exhibits normal scattering envelopes (Fig. 49). The dimensions of the latter sample are close to those of the precursor (Tables 9 and 10).

Table 11 Cross-sectional radius of gyration for sample F3-PS-PS41 S68 for different counterions and for different amount of added salt

Counterion	Added salt	$R_{g,c}$ (nm)
H$^+$	10^{-1} mol L^{-1}	3.94
H$^+$	10^{-2} mol L^{-1}	4.06
H$^+$	10^{-3} mol L^{-1}	4.01
H$^+$	10^{-4} mol L^{-1}	4.16
H$^+$	10^{-5} mol L^{-1}	4.05
H$^+$	"Salt-free"	4.04
Cs$^+$	"Salt-free"	4.14
Ba^{2+}/H$^+$=0.028	"Salt-free"	3.8
Ba^{2+}/H$^+$=0.055	"Salt-free"	3.6
Ba^{2+}/H$^+$=0.083	"Salt-free"	3.6

5.3.2.1
Cross-Sectional Radius of Gyration

The cross-sectional radii of gyration were determined for sample F3-PS-PS41-S68 as function of added NaCl. The data are included in Fig. 40 and confirm the results obtained for sample PM-PVP25-Me90 that the cross-sectional dimensions do not change with added salt. Within experimental error counter-ion exchange from H$^+$ to Cs$^+$ did not influence $R_{g,c}$ significantly (Table 11). Since there is a strong difference in contrast between H$^+$ and Cs$^+$ the counter-ion and the segment density profile across the cylindrical brush are obviously similar for the sulfonated brushes.

In a second series of experiments barium hydroxide was added in order to monitor the effect of bivalent counter-ions on $R_{g,c}$. Since there is a strong binding between the sulfonate group and Ba^{2+} (BaSO$_4$ is insoluble in water) a counter-ion included cross-linking within the cylindrical brush should be observable in dilute solution. A slight decrease of $R_{g,c}$ is observed which, however, hardly exceeds the experimental error (Table 11). Above 10% Ba^{2+} relative to SO$_3^-$ the brushes precipitate.

5.3.2.2
Intermolecular Structure Formation

X-ray scattering measurements on sample F3-PS-PS41-S68 with H$^+$ and Cs$^+$ counter-ions were extended over a large q-regime and for concentrations ranging from 0.5 g L^{-1}≤ c_p≤20 g L^{-1}. Figure 50 shows the scattering curves of the sample with H$^+$ counter-ions at a polyion concentration c_p=2 g L^{-1} in 10^{-2} mol L^{-1} aqueous NaCl solution. The solid line represents a fit obtained by a Fourier transformation that is performed after having identified the cylindrical geometry:

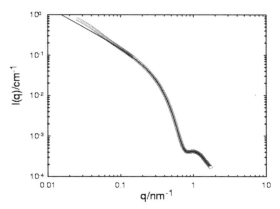

Fig. 50 Log $I(q)$ versus log q for sample F3-PS-PS41-S68 at c_p=2 g L^{-1} and c_s= 10^{-2} mol L^{-1}. The *solid line* represents the scattering curve of a cylinder, as described in the text

$$qI(q) = 2\pi \int\limits_{r=0}^{\infty} P_c(r)J_0(qr)dr \qquad (11)$$

The transformation is performed using the indirect transformation method according to Glatter [130] which includes smoothing of the primary data by a weighted least-square procedure (estimation of the optimum stabilization parameter based on a stability plot) and transformation into real space simultaneously. The fit shown was performed with the data points starting at q=0.06 nm^{-1}, whereas the measured data points start at a value of q=0.026 nm^{-1}. The deviation between the fit and the data at low q-values, as shown in Fig. 50, indicates the deviation of the molecules from cylindrical geometry at low q due to the flexibility of the wormlike brushes. For salt-free solutions scattering peaks appear in the low q-regime at higher concentrations (Fig. 51) whereas the scattering pattern at high q remains the same within experimental error (not shown). The scattering envelopes shown in Fig. 51 contain contributions of the intramolecular scattering, i.e. of the form factor P(q) and of the intermolecular structure factor S(q). Unfortunately it is not possible to extract S(q), even if P(q) is known, because the convolution of P(q) and S(q) is quite complex [94]. Therefore we adopt the relation derived for spherical structures:

$$I(q) = P(q)S(q) \qquad (12)$$

in order to approximately extract S(q) by division of $I(q)$ (interacting rods) by $I(q)\sim$ P(q) of the non interacting rods measured in 10^{-2} mol L^{-1} NaCl solution (Fig. 50). The results are shown in Fig. 52 where for the highest concentrations two maxima are clearly identified. Only for the two lowest

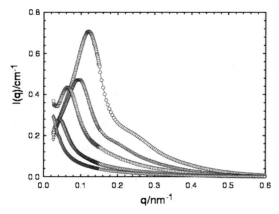

Fig. 51 I(q) versus q for sample F3-PS-PS41-S68 in pure water for different polyion concentration: *triangles*, 1.0 g L^{-1}; *diamonds*, 2.0 g L^{-1}; *squares*, 5.0 g L^{-1}; *inverted triangles*, 10.0 g L^{-1}; *circles*, 20.0 g L^{-1}

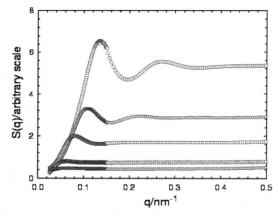

Fig. 52 S (q) as function of q derived as described in text. Symbols as in Fig. 51

concentrations does the intermolecular structure peak move out of the q-window.

The maximum q-values (first maximum) are plotted in Fig. 53 for both, H$^+$ and Cs$^+$ counter-ions. No clear picture evolves as the slope of the sample with Cs$^+$ counter-ions is 0.44 whereas that for H$^+$ counter-ions is 0.34. The dotted line in Fig. 53 would result according to $q^{max} = \frac{2\pi}{<d>} with < d >$ the mean distance of the particles if the particles were evenly distributed throughout the solution, i.e. formed a fcc lattice of spheres. For rod like particles this is only to be expected if the interparticle distance is much larger than the rod length L, i.e. $N_P lt; \frac{1}{L_w^3}$, where N_P is the number concentration. This is shown by the arrow in Fig. 53 assuming a cylinder length per mono-

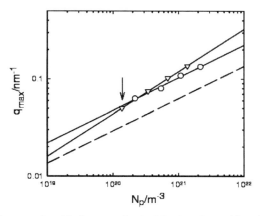

Fig. 53 Log q_1^{max} versus log N_P for sample F3-PS-PS41-S68 with H^+ (*circles*) and Cs^+ (*inverted triangles*) counter-ions. The dashed line represents $q^{max}=2\pi/<d>$ with $<d>$ derived from a uniform particle distribution

Table 12 Peak positions from structure factor maxima, d_s, and mean interparticle distances, $<d>$

Sample	c_P (g L^{-1})	q_1^{max} (nm)	q_2^{max} (nm)	d_s (nm)	$<d>$ (nm)	Peak position ratio
H-PSS	20	0.134	0.27	46.90	87.90	1:2.0
H-PSS	10	0.109	0.22	57.95	110.7	1:2.0
H-PSS	5	0.080	–	78.51	139.5	–
H-PSS	2	0.063	–	99.78	189.4	–
Cs-PSS	20	0.136	0.26	46.2	102.1	1:1.9
Cs-PSS	10	0.102	0.21	61.5	128.6	1:2.0
Cs-PSS	5	0.075	–	84.1	162.0	–
Cs-PSS	2	0.061	–	125.7	219.9	–

mer of 0.2 nm. The measured data lie in the crossover region between the dilute and the semi dilute regime where the scaling exponent is expected to increase from 1/3 to 1/2. Moreover, the mean distance $<d>$ between the particles is significantly larger than that derived from q_1^{max} according to $d_s=2\pi/q_1^{max}$ (Table 12).

Similar discrepancies have been frequently reported for colloidal spheres [131–135] and has led to the postulation of electrostatic attraction resulting in ordered domains of higher local concentration in coexistence with dilute non-ordered particles. For the present system, however, it would be very speculative to assign the slope of 0.44 observed for Cs^+ counter-ions to a stronger attraction as compared to H^+ which could possibly lead to a nematic-like order within the domains. Finally, another strange observation should be mentioned. The ratio q_1^{max}/q_2^{max} of the first and second maximum for the highest concentrations for both, H^+ and Cs^+ counter-ions is equal to

2±0.1 which is normally observed for lamellar ordering. This value is much larger than expected for fcc packing of spheres $(\sqrt{3})$ or for hexagonally packed cylinders $(\sqrt{2})$. One explanation could be that $S(q)$ was derived by application of Eq. (12) which might not be allowed for rod like particles.

5.4
Conclusions on Cylindrical Brushes

Although it may be still a bit preliminary to quantitatively discuss the results obtained so far the following conclusions may be drawn from the investigations described above.

The effective charge density along a single side chain of a cylindrical brush is much less than for a linear flexible polyion. However, the charge density along the contour of the main chain, f_{mc}, is significantly higher than for a linear flexible chain (Table 7). The major difference is that the cylindrical brushes are much thicker ($d\approx7$–10 nm) than typical polymers, i.e. the diameter d is in the order of the Debye screening length κ^{-1}.

Since the fraction of condensed counter-ions is obviously very large it is tempting to estimate κ^{-1} within the volume of a cylindrical brush. For this purpose we assume the diameter of the brushes in solution to be twice the cross-sectional radius of gyration as determined by X-ray scattering. The results are included in Table 7 in terms of the counter-ion condensation and of the inverse Debye screening length. The data indicate that the electrostatic repulsion within a cylindrical polyelectrolyte is very small and probably negligible. However, a strong osmotic swelling due to the Donnan equilibrium should occur which is not at all reflected in the cross-sectional radius of gyration as determined by X-ray scattering.

Neither the main nor the side-chain dimensions of ionic cylindrical brushes differ significantly from their uncharged counterparts. Obviously, the osmotic swelling is not significant for the present small side-chain lengths, i.e. $P_{sc}<50$. The intermolecular structure factor of ionic cylindrical brushes are difficult to interpret. The intermolecular distances derived from the peak of the structure factor is different for H^+ and Cs^+ counter-ions and is always significantly smaller than the mean distance calculated from the concentration of the particles and the known molar mass. Thus, a "two state model" may also be postulated for the present cylindrical polyelectrolyte structures.

6
General Conclusions

Polyelectrolyte brushes constitute a new class of material with very interesting physicochemical properties. The strong stretching of the polymer chains due to segment–segment interactions and electrostatic forces introduce completely new physical properties into monolayers consisting of polyelectrolyte brushes and also into cylindrical brush systems. With the development of new synthetic strategies, simple ways for the preparation of high graft density charged brushes have been paved; these allow, for the first time, a somewhat detailed comparison between theory and the experimental realization of such systems.

For any practical application of polyelectrolyte brushes the influence of multivalent ions, hydrophobic ions, and other polyelectrolyte molecules present in a contacting solution on to the structure of the surface-attached layers or the cylindrical polyelectrolyte brushes are of utmost importance. In particular, study of the interaction of brushes with other polyelectrolyte molecules in solution might open an avenue for the understanding of inter- action of proteins or other charged biomolecules such as DNA, as a special form of charged macromolecules, with charged surfaces. It has become clear that not only the influence of the surface on to the conformation of the pro- tein, but also the influence of the protein on the structure of the polymer layer is important.

However, in general it is a basic requirement for all evaluation of swelling data, and for any practical application, to understand how the brush struc- ture responds to changes in the environment. In same cases the addition of very minute amounts of an certain compound might cause the whole brush to collapse. The strong dimensional changes, which can be observed for sometimes very subtle changes in the chemistry of the environment, could render PEL brushes as very promising materials for sensor and actuator applications.

References

1. Zhao B, Brittain WJ (2000) Progr Polym Sci 25:677
2. Rühe J, Knoll W (2000) Functional Polymer Brushes. In: Ciferri A (ed) Supramolecular polymers, Marcel Dekker
3. Rühe J (2001) Polymer brushes: polymerization to control interfacial properties. In: Encyclopedia of materials: science and technology. Elsevier, The Netherlands, pp 7213– 7218
4. Halperin A, Tirrell M, Lodge TP (1992) Adv Polym Sci 100:31
5. Fleer GJ, Cohen Stuart MA, Scheutjens JMHM, Cosgrove T, Vincent B (1993) Polymers at interfaces. Chapman and Hall, London
6. Alexander S (1977) J Phys 28:977
7. de Gennes PG (1976) J Phys (France) 37:1445
8. de Gennes PG (1980) Macromolecules 13:1069
9. Biesalski M, Rühe J, Kügler R, Knoll W (2002) Polyelectrolytes at solid surfaces. In: Tripathy SK, Kumar J, Nalwa HS (eds) Handbook of polyelectrolytes and their appli- cations. American Scientific Publishers, San Diego, chap 2, p 39
10. Milner ST, Witten TA, Cates ME (1988) Macromolecules 21:2610
11. Milner ST, Witten TA, Cates ME (1989) Macromolecules 22:853
12. Milner ST (1991) Science 251:905
13. Halperin A (1992) Macromol Rep A29:107
14. Skvortsov AM, Gorbunov AA, Pavlushkov VA, Zhulina EB, Borisov OV, Priamitsyn VA (1988) Polym Sci USSR 30:1706
15. Netz RR, Schick M (1998) Macromolecules 31:5105
16. Cosgrove T, Heath T, van Lent B, Leermakers FAM, Scheutjens J (1987) Macromolecules 20:1692

17. Grest GS, Murat M (1994) In: Binder K (ed) Monte Carlo and molecular dynamics simulations in polymer science. Clarendon Press, Oxford
18. Klein J (1996) Annu Rev Mater Sci 26:581
19. Bianco-Peled H, Dori Y, Schneider J, Tirrell M, Sung LP, Satija S (1998) Abstr Pap Am Chem Soc 364, Part 2
20. Bianco-Peled H, Dori Y, Schneider J, Tirrell, M (2001) Langmuir 17:6931
21. Halperin A, Tirrell M, Lodge TP (1992) Adv Polym Sci 100:31
22. Szleifer I, Carignano M (1996) In: Prigogine I, Rice SA (eds) Advances in chemical physics. Wiley, New York
23. Pincus P (1991) Macromolecules 24:2912
24. Borisov OV, Birshtein TM, Zhulina EB (1991) J Phys II (Paris) 1:521
25. Ross RS, Pincus P (1992) Macromolecules 25:2177
26. Zhulina EB, Birshtein TM, Borisov OV (1992) J Phys II (Paris) 2:63
27. Israëls R, Leermakers FAM, Fleer GJ, Zhulina EB (1994) Macromolecules 27:3249
28. Borisov OV, Zhulina EB, Birshtein TM (1994) Macromolecules 27:4795
29. Zhulina EB, Borisov OV (1997) J Chem Phys 107:5952
30. Israëls R, Leermarkers FAM, Fleer GJ (1994) Macromolecules 27:3087
31. Israëls R, Leermarkers FAM, Fleer GJ (1994) Macromolecules 27:3249
32. Fleer GJ (1996) Ber Bunsenges Phys Chem 100:936
33. Lyatskaya YV, Leermarkers FAM, Fleer GJ, Zhulina EB, Birshtein TM (1995) Macromolecules 28:3562
34. Zhulina EB, Birshtein TM, Borisov OV (1995) Macromolecules 28:1491
35. Zhulina EB, Borisov OV (1997) J Chem Phys 107:5952
36. Dünweg B, Stevens MJ, Kremer K (1995) Structure and dynamics of neutral and charged polymer solutions: effects of long-range interactions. In: Binder K (ed) Monte Carlo and molecular dynamics simulations in polymer science. Oxford University Press, New York, p 159
37. Granfeldt MK, Miklavic SJ, Marčelja S, Woodward CE (1990) Macromolecules 23:4760
38. Sjöström L, Åkesson T, Jönsson BJ (1993) J Chem Phys 99:4739
39. Chen H, Zajac R, Chakrabarti A (1996) J Chem Phys 104:1579
40. Lekner J (1991) Physica A 176:524
41. Sperb R (1994) Mol Simulation 13:189; (1998) Mol Simulation 20:179
42. Csajka FS, van der Linden CC, Seidel C (1999) Macromol Symp 146:243
43. Seidel C (2002) Macromolecules, submitted for publication
44. Csajka FS, Seidel C (2000) Macromolecules 33:2728
45. Csajka FS, Netz RR, Seidel C, Joanny JF (2001) Eur Phys J E 4:505
46. Muller F, Fontaine P, Delsanti M, Belloni L, Yang J, Chen YJ, Mays JW, Lesieur P, Tirrell M, Guenoun P (2001) Eur Phys J E 6:109
47. Naji A, Netz RR, Seidel C (2002) Eur Phys J E, submitted for publication
48. Amoskov VM, Pryamitsyn VA (1994) J Chem Soc Faraday Trans 90:889
49. Helm CA (2002) personal communication
50. Fleer GJ, Cohen Stuart MA, Scheutjens JMHM, Cosgrove T, Vincent B (1993) Polymers at interfaces. Chapman and Hall, London
51. Mir Y, Auroy P, Auvray L (1995) Phys Rev Lett 75:2863
52. Tran Y, Auroy P, Lee LT, Stamm M (1999) Phys Rev E 60:6984
53. Tran Y, Auroy P, Lee LT (1999) Macromolecules 32:8952
54. Tran Y, Auroy P (2001) J Am Chem Soc 123:3644
55. Zajac R, Chakrabati A (1995) Phys Rev 52:6536
56. Kopf A, Baschnagel J, Wittmer J, Binder K (1996) Macromolecules 29:1433
57. Lehmann T, Rühe J (1999) In: Adler HPJ, Kuckling D, Jung JC (eds) Macromol Symp 142:1
58. Prucker O, Rühe J (1998) Macromolecules 31:592
59. Prucker O, Rühe J (1998) Macromolecules 31:602
60. Prucker O, Rühe J (1998) Langmuir 14:6893
61. Jordan R, West N, Ulman A, Chou YM, Nuyken O (2001) Macromolecules 34:1606

62. Ejaz M, Tsujii Y, Fukuda T (2001) Polymer 42:6811
63. Biesalski M, Rühe J (1999) Macromolecules 32:2309
64. Biesalski M, Rühe J (2000) Langmuir 16:1943
65. Guo X, Weiss A, Ballauff M (1999) Macromolecules 32:6043
66. Biesalski M, Rühe J (2002) Macromolecules 35:499–507
67. Habicht J, Schmidt M, Rühe J, Johannsmann D (1999) Langmuir 15:2460
68. Biesalski M, Rühe J, Johannsmann D (2000) J Chem Phys 111:7029
68a. One of us (DJ) disagrees with the contents of Ref. 66 and the derivation of the data shown in Fig. 18
70. Gelbert M, Biesalski M, Rühe J, Johannsmann D (2000) Langmuir 16:5774
71. Guo X, Ballauff M (2001) Phys Rev E 64:051406
72. Guo X, Ballauff M (2000) Langmuir 16:8719
73. de Robillard Q, Guo X, Ballauff M (2000) Macromolecules 33:9109
74. Konradi R, Rühe J (2002) PMSE Preprints (in print)
75. Biesalski M, Johannsmann D, Rühe J (2002) J Chem Phys 117:4988
76. Michaeli I (1960) J Polym Sci 48:291
77. Horkay F, Tasaki I, Basser PJ (2001) Biomacromolecules 2:195
78. Schweins R, Huber K (2001) Eur Phys J E 5:117
79. Zhang H, Rühe J, PMSE Preprints (in print)
80. Decher G, Hong JD (1991) Makromol Chem Makromol Symp 46:321
81. Dubas ST, Schlenoff JB (1999) Macromolecules 32:8153
82. van der Steeg HGM, Cohen Stuart MA, de Keizer A, Bijsterbosch BH (1992) Langmuir 8:2538
83. Cohen Stuart MA (1988) J Phys (France) 49:1001
84. Dautzenberg H, Jaeger W, Koetz J, Phillip B, Seidel Ch, Stscherbina D (1994) Polyelectrolyte—formation, characterization and application. Hanser, Munich
85. Marko JF (1994) Europhys Lett 25:239
86. Soga KG, Zuckermann MJ, Guo H (1996) Macromolecules 29:1998
87. Zhulina E, Balazs AC (1996) Macromolecules 29:2667
88. Müller M (2002) Phys Rev E 65:030802(R)
89. Minko S, Müller M, Usov D, Scholl A, Froeck C, Stamm M (2002) Phys Rev Lett 88:035502-1-035502-4
90. Sidorenko A, Minko S, Schenk-Meuser K, Duschner H, Stamm M (1999) Langmuir 15:8349
91. Minko S, Usov D, Goreshnik E, Stamm M (2001) Macromol Rapid Commun 22:206
92. Minko S, Patil S, Datsyuk V, Simon F, Eichhorn K-J, Motornov M, Usov D, Tokarev I, Stamm M (2002) Langmuir 18:289
93. Shusharina NP, Linse P (2001) Eur Phys J E 6:147
94. Mir Y, Auroy P, Auvray L (1995) Phys Rev Lett 75:2863
95. Zajac R, Chakrabarti A (1995) Phys Rev E 52:6536
96. Kopf A, Baschnagel J, Wittmer J, Binder K (1996) Macromolecules 29:1433
97. Tran Y, Auroy P, Lee LT (1999) Macromolecules 32:8952
98. Tran Y, Auroy P, Lee LT, Stamm M (1999) Phys Rev E 60:6984
99. Tran Y, Auroy P (2001) J Am Chem Soc 123:3644
100. Knoll W (1996) Optical properties of polymers. In Encyclopedia of applied physics. VCH, Weinheim, chap 14, p 569
101. Onishi T, Imai N, Oosawa F (1960) J Phys Soc Japan 15:896
102. Oosawa F (1971) Polyelectrolytes. Marcel Dekker, NY
103. Beer M, Schmidt M, Muthukumar M (1997) Macromolecules 30:8375
104. Förster S, Schmidt M (1995) Adv Polym Sci 120:50
105. See also other contributions in this volume
106. Tsukahara Y, Mizuni K, Segawa A, Yamashita Y (1989) Macromolecules 22:1546
107. Tsukahara Y, Tsatsumi K, Yamashida Y, Shimada S (1990) Macromolecules 23:5201
108. Ishizu K, Tsubaki K (1998) Polymer 39:2935

109. Wintermantel M, Schmidt M, Tsukahara Y, Kajiwara K, Kohjiya S (1994) Macromol Rapid Commun 15:279
110. Wintermantel M, Gerle M, Fischer K, Schmidt M, Wataoka I, Urakawa H, Kajiwara K, Tsukahara Y (1996) Macromolecules 29:978
111. Dziezok P, Sheiko SS, Fischer K, Schmidt M, Möller M (1997) Angew Chem 109:2894
112. Sheiko SS, Gerle M, Fischer K, Schmidt M, Möller M (1997) Langmuir 13:5368
113. Gerle M, Fischer K, Roos S, Müller AHE, Schmidt M, Sheiko SS, Prokhorova S, Möller M (1999) Macromolecules 32:2629
114. Djalali R, Hugenberg N, Fischer K, Schmidt M (1999) Macromol Rapid Commun 20:444
115. Ito K, Tanaka H, Imai G, Kawaguchi S, Isuno S (1991) Macromolecules 24:2348
116. Tsubaki K, Ishizu K (2001) Polymer 42:8387
117. Kawaguchi S, Akaike K, Zhang Z-M, Matsumoto H, Ito K (1998) Polym J 30:1004
118. Kawaguchi S, Mauiruzzaman M, Katsuragi K, Matsumoto H, Ito K, Hugenberg N, Schmidt M (2002) Polym J 34:253
119. Beers KL, Gaynor SG, Matyjaszewski K, Sheiko SS, Möller M (1998) Macromolecules 31:9413
120. Börner HG, Beers KL, Matyjaszewski K, Sheiko SS, Möller M (2001) Macromolecules 34:4375
121. Cheng G, Böker A, Zhang M, Krausch G, Müller AHE (2001) Macromolecules 34:6883
122. Deffieux A, Schappacher M (1999) Macromolecules 32:1797
123. Deffieux A, Schappacher M (2000) Macromolecules 33:7371
124. Fischer K, Schmidt M (2001) Macromol Rapid Commun 22:789
125. Sheiko SS, Prokhorova SA, Schmidt U, Dziezok P, Schmidt M, Möller M (2000) In: Tsukruk VV, Wahl KJ (eds) Microstructure and microtribology of polymer surfaces. ACS Symposium Series 741, Washington DC
126. Volk NH (2002) PhD Thesis, University of Mainz
127. Förster S, Schmidt M, Antonietti M (1990) Polymer 31:781
128. Förster S, Schmidt M, Antonietti M (1992) J Phys Chem 96:4008
129. Hugenberg N, Loske S, Müller AHE, Schärtl W, Schmidt M, Simon PFW, Strack A, Wolf BA (2002) J Non-Cryst Solids 307:765
130. Glatter O (1997) J Appl Cryst 10:415
131. Ise N (1986) Angew Chem Int Ed Engl 25:323
132. Ise N, Matsuoka H, Ito K (1989) Macromolecules 22:1
133. Ito K, Yoshida H, Ise N (1994) Science 263:66
134. Yoshida H, Ise N, Hashimoto T (1995) Langmuir 11:2853
135. Gröhn F, Antonietti M (2000) Macromolecules 33:5938

Received January 2003

Adv Polym Sci (2004) 165:151–175
DOI 10.1007/b11269

Lipid and Polyampholyte Monolayers to Study Polyelectrolyte Interactions and Structure at Interfaces

H. Möhwald[1] · H. Menzel[2] · C. A. Helm[3] · M. Stamm[4]

[1] Max-Planck-Institut für Kolloid- und Grenzflächenforschung, Am Mühlenberg 1,
14424 Potsdam, Germany
E-mail: *Helmuth.moehwald@mpikg-golm.mpg.de*
[2] Institut für Technische Chemie, Technische Universität Braunschweig,
Hans-Sommer-Strasse 10, 38106 Braunschweig, Germany
[3] Institut für Physik, Bereich Angewandte Physik, Ernst-Moritz-Arndt-Universität,
Friedrich-Ludwig-Jahn-Strasse 16, 17487 Greifswald, Germany
[4] Institut für Polymerforschung Dresden, Hohe Strasse 6, 01069 Dresden, Germany

Abstract Lipid and polyampholyte monolayers are shown to constitute interesting models for study of electrostatic and steric interactions of polyelectrolytes at interfaces. Measurements with Langmuir monolayers at the air/water interface yield access to the energetics and kinetics of adsorption processes. Newly developed types of optical microscopy, X-ray diffraction, and reflection techniques yield fine structural details, for example lipid structure, segment density profiles, and domain structure. Scanning probe microscopy enables resolution of adsorbate structure on solid surfaces which proves to be most interesting for polyampholytes, for which the charge density can be varied via pH. The results can only partly be explained by theory. Unexpected and not yet explained findings are attractive interactions of polyelectrolytes and interfaces and long-range interactions affecting kinetics and micro phase separations. They can, however, be used as a primary means of controlling interface properties and structure.

Keywords Polyampholytes · Adsorption · Micelle · IEP · Interfaces · Surface Modification

1
Introduction

Langmuir monolayers, the most defined arrangements of amphiphiles at a
water surface, are not only interesting because of various aspects of physics
in two dimensions, they are also excellent model systems for study of many
different problems in materials and biosciences [1]. One of these areas is
polymer science, in which the behavior of polymers at interfaces is a topic of
highest relevance [2]. In this context the hydrophilic monolayer part may be
considered as a surface of variable or defined composition, also with special
charge density, on to which a water-soluble polymer in the subphase may be
adsorbed (Fig. 1). The system has the special advantage that surface energetics can be assessed via surface tension measurements and that adsorption
equilibrium, in the case of lateral mobility, is established in a much shorter
time than with solid surfaces. The system has recently gained more attention
because the last two decades have seen a wealth of new technical developments for study of liquid interfaces at many length scales down to the molecular level [1, 3].

Polyampholytes on the other hand consist of oppositely charged groups
and are a special type of polyelectrolyte. The behavior of polyampholytes is
mainly determined by the net charge and the ratio between the oppositely
charged groups [4]. Natural polyampholytes, like proteins, play a significant
role in the processes of life, and protein adsorption on a solid substrate has
been the topic of numerous publications [5–8]. Whereas proteins have specific compositions and structures, due to their biological function, synthetic
polyampholytes have a much simpler structure. Therefore, synthetic
polyampholytes may serve as model systems for understanding the mechanism of adsorption. Theoretical aspects of the adsorption behavior of
polyampholytes have been discussed recently [9] and diblock polyampholytes were found to form monolayers of micelles [10, 11] similar to those
observed for uncharged amphiphilic diblock copolymers [12].

The present contribution, concentrating on polyelectrolytes at interfaces,
is not meant to report in detail on one system, but should demonstrate what
can be learnt from studying these systems. We thus give illustrative exam-

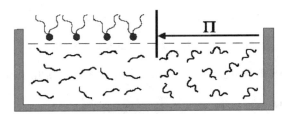

Fig. 1 The model system. An amphiphile, in most cases with an ionic hydrophilic head
group and two aliphatic tails at the air/water interface with easy manipulation of surface
pressure, π, and molecular area, A. Polyelectrolyte dissolved in the subphase may bind
to the monolayer

ples from work of our groups. Also experimental details will be given only as far as needed to understand the data presented.

2
Molecules and Systems

Details of material handling and purification can be deduced from the original publications cited. Here we would like to concentrate on those features which qualified the molecules for selection. Dioctadecylammonium bromide (DDAB) is a cationic double chain surfactant forming stable monolayers with (at room temperature) a phase transition from a liquid expanded (LE) to a liquid condensed (LC) state. In the LC phase the aliphatic tails are known to order but at any pressure the tails are tilted with a tilt angle above 40°. This is probably because of packing constraints in the head group region. The anionic phospholipids dimyristoylphosphatidic acid (DMPA) and dipalmitoylphosphatidic acid (DPPA) form stable monolayers with vertical tails (at high pressure), and DMPA also undergoes a LE/LC phase transition. Cardiolipin comprises basically two DPPA molecules linked in the head group region. At room temperature it has an LC phase with vertically oriented tails, at higher temperatures one may observe an LE/LC transition. The amphiphilic block copolymer polyethylethylene (PEE)–polystyrenesulfonic acid (PSS) is used to graft the anionic PSS at the interface. The side branch of the PEE backbone serves to inhibit condensation, hence the hydrophobic anchor is essentially fluid-like.

The cationic polyelectrolytes polyallylamine hydrochloride (PAH) and polydimethyl diallylammonium chloride (PDADMAC) are known to bind to anionic surfactants. They are distinguished by the repeat length along the backbone. The bulky side group probably gives PDADMAC greater stiffness. The fact that the charge on the amine of PAH depends on pH is not made use of here. Poly(p-phenylene sulfonate) (PPPS; sodium salt) is a rather stiff anionic polyelectrolyte with (compared to PSS) a rather low charge density. The anionic carboxymethylcellulose (CMC; sodium salt) is presumably rather stiff and strongly hydrated. DNA is the anionic polyelectrolyte with the largest persistence length. For this reason it is presented here, but there are obviously many biophysical and biotechnological reasons to study its interaction with membranes.

On the other hand the adsorption of the weak diblock polyampholyte poly(methacrylic acid)-*block*-poly((dimethylamino)ethyl methacrylate), PMAA-*b*-PDMAEMA, with varying copolymer composition and molecular weight has been investigated [13, 14]. It is an interesting model system because of its charge on the two blocks varying with pH. The adsorption was performed on silicon substrates from aqueous solutions at different pH values. The comparison of polyampholytic systems with different copolymer composition and molecular weight provides a good basis to understand adsorption behavior and structure formation at an interface.

3
Polyelectrolyte Adsorption at Monolayer Surfaces

A prerequisite for studies of thermodynamics and kinetics of adsorption is to prepare a macroscopically homogeneous surface and subphase. An elegant method to achieve this and to exchange the subphase in a short time is to use a two-compartment trough where in one trough the polyelectrolyte is dissolved homogeneously in the subphase and the monolayer is spread on a second trough and compressed to a desired molecular area, A. This monolayer is then moved laterally on the first trough at constant A and change in surface pressures or other surface parameters can then be recorded [15]. In the example of Fig. 2 a cationic monolayer is transferred over a subphase containing different anionic polyelectrolytes. Obviously binding of the polyelectrolyte may lead to an increase or a decrease in surface pressure, depending on the polyelectrolyte, and also the slopes of the curves may deviate. They either change continuously (CMC, PSS) or there may be an inflection (PPPS). This will be discussed below. The sign of the changes can be understood qualitatively by inspecting the pressure/area isotherms of Fig. 3. They show a change in slope at a molecular area near 0.8 nm^2 corresponding to a first-order phase transition from a liquid expanded (LE) to a liquid condensed (LC) phase [16, 17]. The condensation of the film with PSS binding to it (curve b) can be understood as screening of the Coulombic repulsion of the charged head groups due to the anion binding. This electrostatic contribution to the surface pressure has been calculated on the basis of Gouy–Chapman theory and may amount to about 10 mN m^{-1} [18]. More difficult to quantify is the pressure increase upon polymer binding. It may be due to:

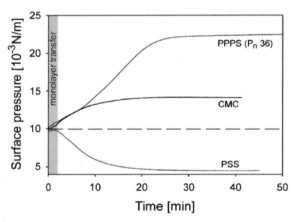

Fig. 2 Surface pressure as a function of time after depositing a DODAB monolayer at fixed film area on a subphase containing the anionic polyelectrolytes PSS, CMC, or PPPS. Subphase polyelectrolyte concentration 5×10^{-5} monoMol

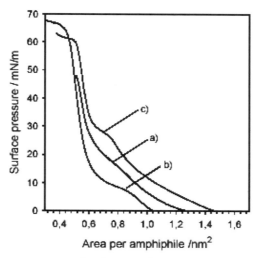

Fig. 3 Surface pressure versus molecular area for a DODAB monolayer on pure water (*a*), on a 5×10^{-5} monoMol PSS solution (*b*), and on a 5×10^{-5} monoMol CMC solution (*c*)

1. steric constraints of the polymer/lipid counter-ion binding, e.g. if the two-dimensional charge density of the polyelectrolyte projecting on the surface is higher than that of the lipid monolayer at the same pressure;
2. entropic constraints, since the polyelectrolyte may favor a disordered configuration with lower lateral charge density. The latter may be less pronounced for stiff polyelectrolytes like DNA or PPPS. For longer chains entropic repulsion dominates, and therefore one observes a lateral expansion for high molecular weight PSS adsorption in contrast to the finding for short PSS (below) [19].
3. insertion of fractions of the polyelectrolyte into the hydrophilic or hydrophobic region of the monolayer, since typical polyelectrolytes also possess hydrophobic moieties lending them some amphiphilicity. The latter effect obviously is not important at high pressures since all pressure/area isotherms before collapse tend towards a limiting molecular area around 0.5 nm^2. This is larger than the area of 0.4 nm^2 expected for two aliphatic tails oriented normal to the surface and indicates a chain tilt. The arguments above to explain monolayer condensation or expansion upon polyelectrolyte coupling also explain the shift in the pressures corresponding to the phase transition since they can likewise be used to favor or disfavor the LC phase with respect to the LE phase.

It is most convenient to use the change in surface pressure, $\Delta\pi$, for a fixed film area as a measure of polyelectrolyte density A_p within the film. A linear $\Delta\pi$ versus A_p relation, however, is only expected if the lipid undergoes no phase transition and if the polymer also does not rearrange cooperative-ly. The first condition can be verified from pressure/area isotherms, the second one also, but it will later prove even more instructive to learn about these transitions.

If adsorption is diffusion-controlled we expect the well-known dependence of A_p on time t, diffusion coefficient D and subphase polymer concentration c_o [20, 21]:

$$A_p(t) = \left(\frac{2}{\sqrt{\pi}}\right) c_o \sqrt{D \cdot t} \tag{1}$$

For low A_p one expects the square root dependence on time to hold; for high A_p there is interaction between adsorbed polymers and also the $A_p \Delta \pi$ relationship is not expected to be linear. Therefore we refrain here from fitting the measured time dependence but define an easily measurable time $t_{1/2}$ corresponding $t_{1/2}=t$ at which $\frac{\Delta\pi}{\Delta\pi_{max}} = 0.5$ where half of the total pressure change has occurred. This time should correspond to half of the total polymer adsorption. This is true only in rough approximation, but the definition is useful for comparing systems and conditions. From Eq. (1) one thus derives

$$t_{1/2} = \left(\frac{A_{p(1/2)}^2}{4D} \cdot \pi\right) c_o^{-2} \tag{2}$$

For dilute polymer solutions one expects D to depend on concentration according to $D \approx c_o^{-h}$ with $-0.1 > h > -1$, and one thus expects a relationship [22]:

$$t_{1/2} \approx c_o^{-2-h} \tag{3}$$

The double logarithmic plot according to Eq. (3) in Fig. 4, however, yields values of h of approximately -1.4, i.e. considerably larger in absolute value than expected from diffusion measurements in solution [23]. We have no final explanation for this discrepancy. One reason may be that for charged species and charged interfaces the simple diffusion model leading to Eq. (1) is insufficient—long range attraction from the charged interface leads to directed transfer, and the polyelectrolyte concentration near the interface is increased [24]. This calls for numerical calculations of adsorption kinetics in a way similar to that pursued for the adsorption of low molecular weight ionic surfactants [25].

The more complex adsorption kinetics for PPPS in Fig. 2 suggests that it cannot be described by a simple diffusion model. The sigmoidal shape also indicates that cooperative processes have to be taken into account, and these may be rate-limiting. In the case of a rigid polymer like PPPS one may assume a nematic ordering of the rod-like polymers in two-dimensional sheets and the pressure change $\frac{\Delta\pi}{\pi_{max}}$ may reflect the amount $U(t)$ of polymers in

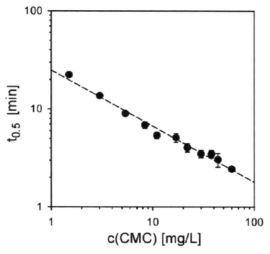

Fig. 4 $t_{1/2}$ as a function of CMC subphase concentration on a double logarithmic scale

these sheets. Polymer crystallization is frequently described by the semi-empirical Avrami equation [26]

$$U(t) = 1 - \exp(-kt^n) \qquad (4)$$

where k is a rate constant and n an exponent characterizing the growth model. For two-dimensional growth one expects $n = 3$, whereas if growth is limited by diffusion from a half-sphere below the "crystalline" sheet n reduces to 2. Hence in real systems one expects n between 2 and 3 and this is indeed found from simulations [27]. Figure 5 demonstrates that the pressure

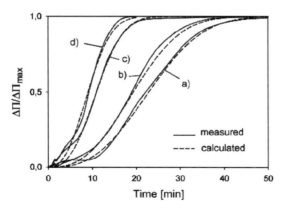

Fig. 5 Measured and calculated (using the Avrami equation, Eq. 4) time dependence of the pressure change $\Delta\pi / \pi_{max}$ for a DODAB monolayer varying the PPPS concentration in the subphase: 10^{-5} monoMol (a), 2×10^{-5} monoMol (b), $4\cdot10^{-5}$ monoMol (c), 8×10^{-5} monoMol (d)

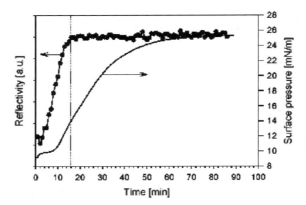

Fig. 6 Optical reflectivity ΔR at 325 nm from DODAB adsorbing at the interface and change in surface pressure as a function of time after transfer on PPPS- (1.4×10^{-5} mono-Mol) containing subphase

change can be simulated reasonably well with the Avrami equation. Discrepancies when t is small may be explained by distortion from monolayer transfer and by induction of crystal nucleation.

The assumption underlying Eq. (4) is that crystals are formed at the interface and it is highly desirable to verify the existence of these crystals. For another rod-like polyelectrolyte, DNA, we will demonstrate the existence of organized rods by surface diffraction (see below). Principally one can also prove local order by use of different types of optical microscopy, for example Brewster angle microscopy [28, 29], a special version of imaging ellipsometry [30], fluorescence [31], or scattering microscopy [32]. As an example of a quantitative measurement Fig. 6 compares reflectivity ΔR and surface pressure as a function of time for the system of Fig. 5 [33]. In this case ΔR is measured at a wavelength (325 nm) corresponding to the polyelectrolyte absorption maximum and therefore measures polyelectrolyte concentration at the interface. One thus deduces from Fig. 6 that adsorption saturates after 16 min, but the surface pressure change requires four times as long. This supports the assumption made for the simulations of Fig. 5 that in this case we are not simulating polyelectrolyte adsorption but its ordering which is then reflected in a pressure change.

4
Polyelectrolyte Arrangement at Interfaces

Many theoretical efforts have been made to describe the arrangement of polymer segments at interfaces and the corresponding energetics [34–36]. The extremes are a pancake–or a brush like arrangement–but subtleties like ion condensation within the brush or detailed segment density distribution are also important for understanding the properties. Reflectivity measurements have proved most valuable to learn about segment distribution, and

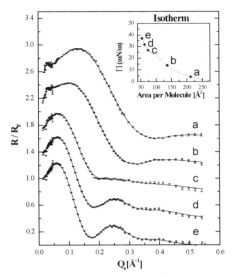

Fig. 7 X-ray reflectivity R normalized to the Fresnel reflectivity R_F expected for an infinitely sharp interface for a DODAB monolayer on a subphase containing 1 mol L^{-1} CsCl and 10^{-3} monoMol PSS at different pressures indicated in the isotherm in the inset. For clarity the reflectivity curves are displaced by 0.4 units

among these neutron reflectivity has gained outstanding relevance [37]. This is because one may deuterate segments completely or partly and thus gain specific contrast.

However, due to limited flux one cannot measure at high Q-values as desirable, measurement times are long and this limits systematic parameter variation. An alternative could be X-ray reflection measurements because even laboratory sources provide high flux and synchrotrons provide ample intensity. However, in most cases scattering relies on electron density contrast, and in this respect many polymers are not distinguished much from water. A new development has been to increase the contrast via insertion of heavy counter-ions into the subphase [38, 39]. In the example of Fig. 7 the X-ray reflectivity of a DODAB monolayer with 1 mol L^{-1} CsCl and 10^{-3} monoMol PSS in the subphase is given as a function of wave vector transfer $Q_z = \frac{4\pi}{\lambda}\sin\varphi$ where φ is the incidence angle and λ the wavelength [19]. The reflectivity R divided by the reflectivity R_F expected for an infinitely sharp interface is given according to [40]:

$$\frac{R}{R_F} = |\frac{1}{\rho_{sub}} \int \frac{d\rho}{dz} \exp(iQ_z Z) dz|^2 \tag{5}$$

To derive the electron density profile $\rho(z)$ one has to use models; the most successful have been slab models to describe the profile. For the system of Fig. 7 one basically has to use a slab containing the aliphatic tails, a thin and

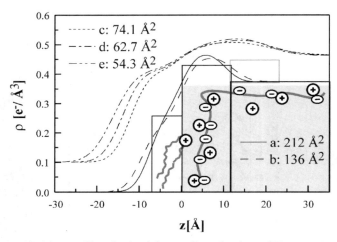

Fig. 8 Electron density profiles obtained from a fit to the data of Fig. 7. $z=0$ corresponds to the alkyl tail/head group interface. For clarity the profiles d–f of the condensed phase are vertically shifted by $0.1e^-/Å^3$. Also shown is a schematic of the molecular arrangement within the three-slab model. The thick and diffuse slab to the right extends beyond the scale of the figure

dense one near the head groups and a less dense and elongated slab furthest into the subphase [19]. The density of the two hydrophilic slabs is largely influenced by the Cs^+ ions which are very close to the anionic polyelectrolyte, and the electron density therefore reflects the segment density distribution. Of course one has to verify that the structure is not changed ion-specifically. This has been done by comparing pressure/area isotherms with CsCl and NaCl at different concentrations in the subphase and also comparing X-ray reflectivity as far as meaningful data could also be extracted with NaCl in the subphase.

The most surprising finding is the enrichment of polyelectrolyte segments in a thin slab of about 1 nm thickness near the head groups as sketched in Fig. 8. A careful analysis revealed that the binding is far from stoichiometric, i.e. the ratio of PSS monomers and lipid head is about 1.2 at highest compression and increases toward 5 on film expansion. This proves that the interaction polyelectrolyte/monolayer is not only electrostatic in nature. On the contrary, one observes overcharging, hence binding occurs even against electrostatic repulsion. This may be explained by at least two different effects:

1. A charged molecule in high dielectric medium (water) experiences image forces at an interface towards low dielectric medium (hydrocarbon, air) [41]. This favors counter-ion binding, hence the polyelectrolyte is not charged near the interface.

2. The hydrophobic moieties of PSS cause attraction to the interface. This is most plausible, since one also observes PSS adsorption at a free water surface [42].

The second hydrophilic slab extends up to 8 nm and contains a much lower segment density (0.1 to 1 per lipid). It is not observed for low-molecular-weight PSS (M_w 4300 Da) and is hardly observed for low salt. Its thickness decreases on film compression toward 2 nm. Qualitatively this thickness decrease can be understood in terms of residual electrostatic interactions between the surface and the polyelectrolyte tails—on compression surface overcharging is reduced and hence the lowered repulsion (from head+PSS in the first slab) enables condensation of the remaining segments in the adjacent slab.

We are not aware of theoretical predictions for the technologically most important high-salt regime. Still these results confirm charge reversal upon polyelectrolyte binding which has been observed in the fabrication of polyelectrolyte multilayers. It is also obvious that the complexity of electrostatic conditions at the interface and their change upon adsorption will cause complicated adsorption kinetics, as we have discussed above.

An alternative way of grafting a polyelectrolyte at an interface is by using an amphiphilic block copolymer [38, 39]. In the example to be discussed a fluid hydrophobic block serves as an anchor at the air/water interface and the polyelectrolyte block is immersed into water. In this case the two-dimensional polyelectrolyte density can be controlled in a defined way and varied via film compression. Careful analysis of the pressure/area isotherms reveals changes in the slope indicating higher-order phase transitions. Fortunately the X-ray reflectivity exhibits pronounced structure, and therefore detailed information can be deduced. The pronounced minimum at large Q_z in Fig. 8 is dominated by interference of a wave reflected at the hydrocarbon/air interface and from the center of the middle slab. Its shift upon compression is therefore expected, because the thickness of the hydrocarbon slab increases (in this case from 0.8 nm to 2.4 nm). The maximum and shoulder for $Q_z < 0.1$ Å^{-1} are dominated by the extension of the polyelectrolyte into the subphase. The insensitivity of the Q_z position with respect to lateral density indicates that in this case the profile extension does not change much. However, the density changes, as is revealed, e.g., from a change in maximum reflection intensity. Profiles deduced from fitting show this in detail at the bottom of Fig. 8. One again realizes the increased PSS density in a 1 nm layer beneath the hydrocarbon slab. To describe satisfactorily the decay of density up to 15 nm into the subphase two additional PSS slabs had to be introduced, and it is most instructive to compare the total lengths of the three PSS containing slabs. This is given in Fig. 9 (top) as a function of area per grafted polymer for various ionic strengths. For low salt (1 mmol L^{-1}) one observes the largest length (12.5 nm), irrespective of molecular areas. This is expected for an osmotic brush where the counter-ions are embedded within the volume defined by $A \times H$, where A and H are, respectively, the area and

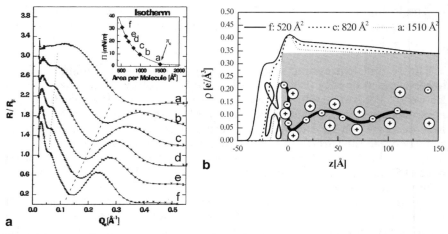

Fig. 9 (a) Normalized X-ray reflectivity from the block copolymer PEE$_{114}$ PSS$_{83}$ on a subphase containing 0.5 mol L^{-1} KCl measured along the isotherm given in the insert. The full lines correspond to a fit of the electron density profiles in **b** to the data. For clarity the reflectivity curves are shifted by 0.5 units, (b) Electron density profiles and corresponding arrangement of hydrophobic and hydrophilic block

length of the brush. The osmotic pressure p then is given by the density of the counter-ions $\frac{Q}{A \cdot H} \approx p$ [43]. This yields an osmotic force $f_{osm} = p \cdot A = \frac{Q}{H}$

f_{osm} is balanced by an elastic force $f_{el} \approx \frac{H}{a^2 \cdot N}$ where a is a segment length and N the number of segments. The balance then yields $H \cong a\ (QN)$, irrespective of molecular area.

Increasing the salt concentration one expects the lateral interactions to be reduced and the polyelectrolyte to be less stretched. This is indeed observed. Also as typical for the salted brush the brush length increases upon compression [44]. For this case one theoretically expects the dependence on density as depicted by the full line describing the measurements for 0.1 mol L^{-1} salt. At still higher salt contents one expects counter-ion condensation [45], hence part of the counter-ions do not contribute to the osmotic pressure. In this case one expects a further shrinking of the brush with salt content. In the extreme of a collapsed polymer one expects the length to be inversely proportional to the molecular area which corresponds to the steepest slope at highest salt content.

In summary with this system it has been possible to observe the transition from the osmotic to a salted and then to a collapsed brush. One might suspect that these transitions are also reflected in the pressure/area isotherms. However, we have not yet been able to relate the breaks in the isotherm slopes with parameters deduced from X-ray reflectivity. This would require an even higher precision of data analysis, and precision in density determination better than 1% is difficult to achieve. We realize that the transition at π_c is accompanied by a change in relaxation times as expected for a glass transition, but also as expected for three dimensional glasses there is no discontinuity in any structural parameter.

We should also note that the finding of enrichment near the head groups is not unique for PSS. It has also been observed for polyampholytes like gelatin [46].

5
Influence of Polyelectrolyte on Monolayer Structure

Since it is known that lipid monolayers exhibit a richness of phases in a narrow temperature and pressure range their structure is expected to be a very sensitive measure of interactions at interfaces. For ordered arrangements X-ray diffraction is known to be the most direct and precise technique for structure analysis, and these measurements have become possible in the last 15 years by making use of the high brilliance of synchrotron sources [47]. With this technique an X-ray beam strikes the surface at grazing incidence and if the electron density projected on the surface is laterally periodical one observes Bragg peaks indicating the scattered intensity as a function of the scattering angle, α. The wave vector transfer Q_{xy} is defined as $Q_{xy} = \frac{2\pi}{\lambda}\sin\alpha$ where λ is the X-ray wavelength. Because the signal is obtained by averaging over many randomly distributed domains one basically observes a two-dimensional powder pattern. It is important to note that up to now no signals could be ascribed to head groups, i.e. one basically measures the arrangement of aliphatic tails. If the tails are uniformly tilted the maximum of intensity is moved out of the horizontal towards a wave rector, Q_z^0, and this is related to the tilt angle, t, by $\tan t = \frac{Q_z^o}{Q_{xy}}$ [48].

It turns out that this technique enables measurement of even small tilt angles, in favorable cases with precision close to $1°$. As an example Fig. 10 compares contour plots measured for the phospholipid DPPA at different pressures on pure water and on a polyelectrolyte (PDADMAC) solution [49]. At low pressures (upper left) on water one observes three diffraction spots with $Q_z^0 > 0$. This means the lattice is oblique and the tails are tilted. Increasing the pressure one observes two diffraction spots, one with $Q_z^0 > 0$, showing that the lattice is rectangular with tilt relative to one lattice plane. On further increasing the pressure there is only one peak with $Q_z^0 = 0$, i.e. the lattice is hexagonal and the chains are oriented normal to the plane. In contrast, with PDADMAC in the subphase one observes at all pressures a rectangular lattice with chain tilt and the tilt angle (i.e. $Q°_z$) decreases only slightly with pressure.

The tilt angle derived from a detailed analysis of the peaks as a function of pressure is given in Fig. 11. In the absence of polymer one observes the reduction of t to zero whereas for PDADMAC $t > 25°$ even at the highest pressure. For comparison we also include the tilt angle measured for PAH in the subphase. The binding of this cationic polyelectrolyte apparently has only a minor influence on monolayer structure. The polyelectrolyte coupling may be best understood looking at the model of Fig. 12 which should hold specif-

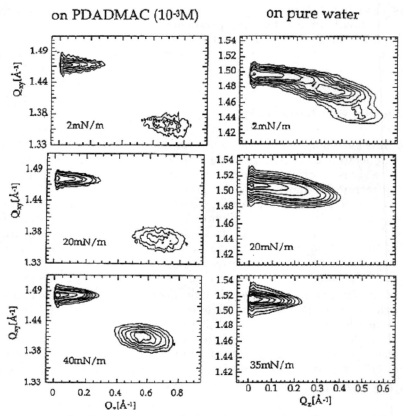

Fig. 10 Intensity distribution (contour plots) in the Q_{xy}, Q_z plane of grazing incidence X-ray diffraction for a monolayer of DPPA at different pressures on pure water (left) and on a 10^{-3} monoMol PDADMAC solution. The circles indicate peaks of maximum intensity

ically for PDADMAC. The insensitivity of the structure with respect to pressure suggests that the polymer at the interface is stiff and tightly coupled to the lipid. The rationale for this model is that the lipid tails form a centered rectangular lattice with tilt along axis a. The axis normal to a is 7.6 Å long, i.e. twice the PDADMAC repeat distance. Thus if PDADMAC is oriented perpendicular to a it is commensurate with the lipid lattice and one obtains 1:1 binding stoichiometry. The model further explains the following findings [49]:

– The diffraction spot for the periodicity along the polymer rod does not change at all with pressure. It is also narrower than in the absence of polymer, i.e. the positional correlation along this direction is increased by polymer coupling.

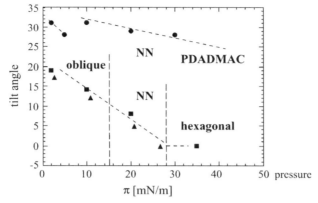

Fig. 11 Tilt angle derived from analysis of diffraction spots (Bragg rods) versus surface pressure for a DPPA monolayer on pure water (*squares*), on 10^{-3} monoMol PDADMAC (*circles*) and on 10^{-3} monoMol PAH(*triangles*)

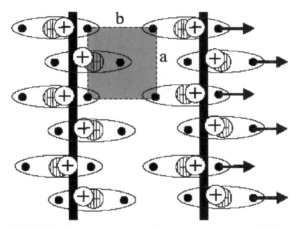

Fig. 12 Model of PDADMAC arrangement (linear rod) and a DPPA monolayer (top view). The phospholipid exhibits two aliphatic tails (*black dots*) that are uniformly tilted (*arrows*) along lattice vector a. Lattice spacing, **a**, corresponds to twice the repeat length of PDADMAC, **b**, to the polymer diameter

– Also along b the lattice cannot be compressed to obtain a smaller tilt angle since from molecular models the polymer diameter is close to b. One would thus compress against lateral steric repulsion of the polymer.

Although the model explains all findings we hesitate to call PDADMAC a stiff rod. Still one has to be aware that binding to the interface stiffens the polymer and that the model explains findings with X-ray concerning posi-

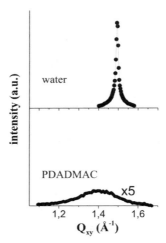

Fig. 13a In-plane diffraction intensity for a cardiolipin monolayer on water without and with PDADMAC in the subphase

tional correlation lengths which are typically 30 repeat units. It is also true that this model describes an extreme situation:

- With increasing ionic strength the coupling is reduced and hence the lipid monolayer becomes more compressible [50].
- The latter is also valid for binding of a copolymer of PDADMAC and the non-ionic polymer [51].

With the model of Fig. 12 we can also explain the opposite finding with PAH in the subphase. The repeat distance is 2.6 Å and one does not find any low-order spacing that would be commensurate with it. Also from the molecular structure one expects the polymer to be more flexible. Hence we do not expect PAH to induce order by binding. On the other hand we would have expected that it reduces head group repulsion thus condensing the film. Apparently this is counteracted by entropic disorder induced by the polymer.

The other extreme where polyelectrolyte coupling creates disorder is given in Fig. 13a. There the monolayer is of cardiolipin, a lipid existing of two coupled diacylphosphatidic acids [52]. The pressure/area isotherms indicate that it exhibits a liquid condensed phase, and this is confirmed by the sharp diffraction line. With PDADMAC in the subphase the film is expanded, and one observes only a very broad diffraction line. From the width one deduces a correlation length of about two spacings, indicating liquid-like order. Still in both cases one may determine the area per molecule from the X-ray spacings, and these are included in the isotherm graphs. One realizes that the point lies on the isotherm in the absence, but not in the presence of polyelectrolyte. This indicates that the film is heterogeneous with the condensed

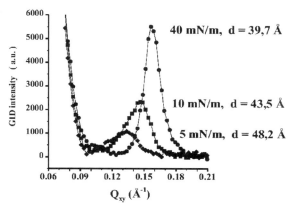

Fig. 13b In-plane diffraction intensity versus wave vector transfer Q_{xy} for DNA binding to a DODAB monolayer at high surface pressure

and more ordered region observed by means of X-rays and a disordered region not measured by diffraction. This is a not unusual problem. Still one should keep in mind that the other extreme, binding causing all sorts of structures, is not observed. This would not yield well-defined diffraction spots.

Up to recently all models on a regular polyelectrolyte arrangement at the monolayer surface could be questioned because there was no direct proof of this. The situation meanwhile has changed since, at least for DNA, we have observed diffraction which can be ascribed to the polymer [53]. As an example of this Fig. 13b gives a diffraction spot corresponding to a repeat distance between rod-like molecules at the interface. In this case the charge density of DNA projected on the surface ($1e^-/0.6$ nm^2) is larger than the charge density of the lipid. Hence for charge compensation one expects a fraction of another layer underneath and the layer near the lipid might be as dense as possible. This is correct only to a certain extent since there is also lateral repulsion between DNA molecules and the spacing can be varied with lateral pressure. This is indeed observed.

6
Polyampholyte Adsorption and Structure at a Solid Interface

The adsorption behavior of weak polyampholytes from aqueous solutions on to a solid wall is strongly determined by the pH of the polymer solution. So it is absolutely necessary to investigate the adsorption as function of pH. As an example the adsorbed amount of two polyampholytes with medium molecular weight of around 60,000 g mol^{-1} and different block sizes is shown in Fig. 14 [13]. In every case the isoelectric point IEP of the polymer, which is determined by the block ratio, is the prominent feature of adsorp-

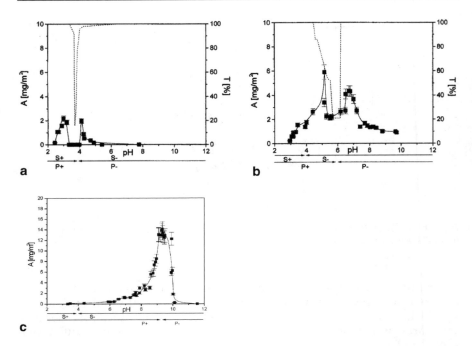

Fig. 14 The adsorbed amount, A, of a polyampholyte as a function of the pH of the adsorption solution. The *solid* and the *dashed* lines are shown as a guide for the eye. The transmission, T, of the solutions (*dotted line*) is shown as a function of pH. The arrows below the graph indicate where the silicon surface, S, and the polyampholyte, P, are carrying a positive or negative net charge. Characteristics of polymers (B1, B3, B4 - **a, b, c**) are: M_n 68,000 g mol^{-1}, 62,000 g mol^{-1}, and 62,000 g mol^{-1}, PMAA/PDMAEMA weight ratios 90/10, 55/45 and 19/81, pH$_{IEP}$ 3.8, 5.9 and 9.3, respectively

tion and solution behavior. The amount of the polymers adsorbed increases from higher and lower pH values towards pH$_{IEP}$, while directly at the IEP a minimum of adsorption occurs (Figs. 14a and 14b). The solution behavior of these polyampholytes at the IEP is characterized by precipitation of the polymer, as indicated by a decrease in the transmission of laser light through the polymer solution. The precipitation at the IEP of the polyampholyte is typical for weak polyampholytes containing an equal amount of positively and negatively charged functional groups [4, 54].

This adsorption regime, including maxima and minima, is explainable in terms of two trends determining the adsorption. The increase of adsorbed polymer is caused by a decrease in net charge of the polyampholyte towards the IEP. Less net charge means a reduction in electrostatic repulsion between the polyampholyte chains, so adsorption at higher density is possible, which leads to a greater amount adsorbed [5, 11, 55]. The minimum directly at the IEP is caused by the second trend. Directly at the IEP the charge of the polyampholyte is shielded by intra- and interchain interactions. So the attractive interactions between the anchor block of the polyampholyte and the negatively charged silicon substrate are reduced and the amount adsorbed decreased. The positive or negative net charge on silicon substrate and polyampholyte copolymer changes with pH and is also indicated in Fig. 14.

For the polymers the maxima and minima in the adsorbed amount are shifted with the IEP of the polyampholytes, which is determined by the block ratio. Furthermore the amount adsorbed increases with the larger positively charged PDMAEMA block, which acts above the pH_{IEP}=3.8 of the silicon wafer as the positive anchor block [9, 56]. The larger PDMAEMA block works in two ways. On the one hand a larger positive block means more electrostatic interaction with the negative silicon surface. On the other hand with the increase of the PDMAEMA block also a shift of the IEP to higher pH values takes place. So the maximum in adsorption occurs at higher pH, where also the silicon surface contains more negative charges.

Both adsorption regimes result in different lateral structures formed by the adsorbed (and dried) polyampholyte on the silicon substrate. The lateral structures of the adsorbed polymers are characterized by differently sized agglomerates [13]. The smallest agglomerates have approximately the size of polyampholyte micelles determined by dynamic light scattering in aqueous solution [55, 56]. The size of polyampholytic agglomerates in aqueous solutions is strongly influenced by pH, which determines the charge of the dissolved polyampholyte. The appearance of some agglomerates is shown in Fig. 15. For the first polymer (B1) at pH=2.8, structures in solution with diameter 90 nm and at pH=4.8 with diameter 132 nm have been observed by light scattering. A polyampholyte of different composition (B3) shows, at pH=6.4, agglomerates with a diameter of 42 nm and at pH=9.1 with a diameter of 96 nm. While, partially, the adsorption of whole micelles can be assumed, larger aggregates observed especially close to the maximum in adsorption near the IEP may be explained by the formation and adsorption of bigger precipitated agglomerates. So the stability of micelles at the interface should be larger than for the other polyampholytes with smaller PDMAEMA block. One may compare those findings with other work in the literature about micelle formation by amphiphilic diblock polymers at an interface, when the adsorbing anchor block is much larger than the non-adsorbing block [57].

Thus, two different adsorption regimes have been identified for PMAA-b-PDMAEMA. The first regime observed with most polyampholytes is determined by two absorption maxima close to the isoelectric point of the polyampholyte and one minimum of the adsorbed amount directly at the IEP. The second regime is observed only for polyampholytes containing a large positively charged PDMAEMA-block and is characterized by one maximum directly at the IEP (Fig. 14c). The explanation for both regimes is given by two influences of electrostatic interactions between polymers and the substrate. On the one hand the repulsive electrostatic interactions between the polymer chains decrease with the polymer net charge towards

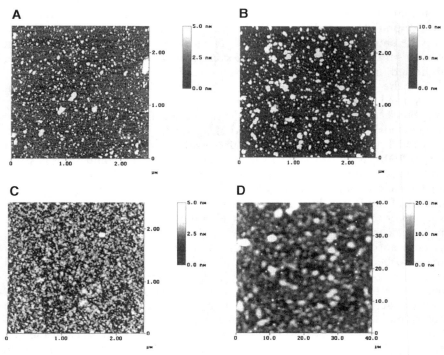

Fig. 15 SFM images of the dry adsorbed polyampholytes from Fig. 14 on silicon substrates. The polymers were adsorbed from solutions of different pH. The adsorbed amount, A, measured by ellipsometry is given. (**A**) Polyampholyte B1 pH=2.7; $A=1.1$ mg m^{-2}; (**B**) Polyampholyte B1 pH=3.0; $A=2.4$ mg m^{-2}; (**C**) Polyampholyte B3 pH=6.9; $A=5.5$ mg m^{-2}; (**D**) Polyampholyte B3 pH=7.5; $A=16.4$ mg m^{-2}

the IEP of the polyampholyte, so the polymer can adsorb at higher density. This tendency increases the amount adsorbed. On the other hand a decrease of net charge of the polyampholyte means also a decrease in the attractive electrostatic interactions with respect to the charged substrate surface. An increase in the positively charged PDMAEMA block causes an increase in the amount adsorbed. This effect was explained by the shift of the IEP to higher pH values and the accompanied increase of attractive interactions between polyampholyte and silicon surface. A clear predominance of the positively charged block leads to the adsorption of homogeneously sized micelles. Under favorable conditions highly regular polyampholyte structures can be observed adsorbed at the substrate [14]. An example is shown in Fig. 16 where from AFM studies and X-ray scattering investigations at grazing incidence a dominant structural size of 64–179 nm is determined, depending on pH during adsorption. In this case structural sizes do not depend strongly on pH and are significantly larger only directly at the IEP.

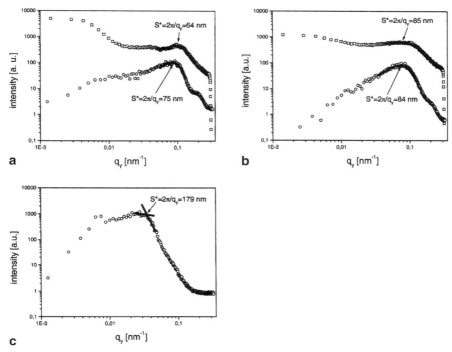

Fig. 16 Power spectral density (PSD) from AFM data (*circles*) and intensity of out of plane scan from X-ray scattering at grazing incidence (*squares*) from an adsorbed diblock polyampholyte (M_n 15,000 g mol^{-1}, PMAA/PDMAEMA weight ratio 33/67) of highly regular lateral structure adsorbed at different pH on a plane silicon wafer: (**a**) 6.1, (**b**) 9.4, and (**c**) 8.5. The most prominent length scale is marked with an arrow

Dependence of the adsorption behavior on the molecular weight has been observed only in the case of polyampholytes of small molecular weight [13]. These polymers show an increase in the adsorbed amount compared to larger polyampholytes, as shown in Fig. 17. This tendency is expected to result from the adsorption of whole micelles of high density. One generally has to consider micelle formation in solution and at the interface to obtain a better understanding of the complex adsorption behavior of polyampholytes.

Other investigations have been performed where silicon substrates were modified from alkaline to acidic and hydrophobic nature [58], where salt concentration was varied during adsorption [55] and where the nature of one of the blocks was changed [59]. Monolayers of polyampholyte micelles were used to modify protein adsorption [60]. An example is shown in Fig. 18 where the adsorption of fibrinogen is controlled by preadsorption of polyampholyte micelles and the phosphate buffer concentration.

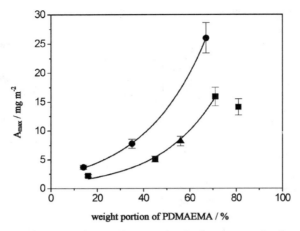

Fig. 17 Comparison between the maximum adsorbed amount of polyampholyte with two high molecular weights and one low molecular weight as function of the block ratio: $M_n{\sim}65{,}000$ g mol^{-1}, *squares*; $M_n=101{,}000$ g mol^{-1}, *triangles*; $M_n{\sim}14{,}500$ g mol^{-1}, *circles*

Blends of homopolyelectrolytes were compared with corresponding diblock polyampholytes with respect to their adsorption behavior [61]. All those investigations show that the use of diblock polyampholytes offers a unique way to control adsorption behavior by pH and to generate structured monolayers with partial highly regular order which can be used to modify the surface properties of solid substrates.

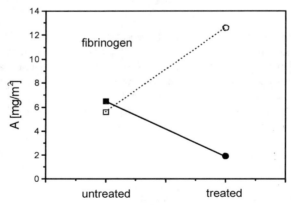

Fig. 18 Adsorption of the protein fibrinogen as influenced by preadsorption of polyampholyte micelles (PMAA/PDAEMA 33/67 weight ratio, $M_n{\sim}15{,}000$) in presence of 0.01 mol L^{-1} (*filled symbols*) and 0.001 mol L^{-1} (*empty symbols*) phosphate buffer solution

7
Conclusion

We have shown here that amphiphile monolayers are an excellent model system to study polyelectrolyte/lipid interactions. Techniques have been developed to study adsorption kinetics and structure at the molecular level. Adsorption kinetics can be described qualitatively with established models but the complications due to long range electrostatic forces and cooperative processes during adsorption keep us far away from a quantitative understanding. Considerable progress has been achieved to elucidate polyelectrolyte arrangement at interfaces and to correlate with theoretical calculations. The lipid monolayer structure, due to its high sensitivity to environmental changes, can be used as a fine sensor to study interfacial interactions. This is needed on one hand to understand these interactions, on the other hand to build-up systems of technical interest or for pharmaceutical applications.

The adsorption of weak diblock polyampholytes at a solid substrate can, on the other hand, be understood qualitatively–amount adsorbed and lateral structures are largely determined by micelle formation in solution, which is strongly influenced by pH. In this way quite different and, in part, highly regular structures at the surface may be generated, depending on copolymer composition, molecular weight, pH, and salt concentration of the solution. This parameter field offers several possibilities and resulting structured polyampholyte monolayers may be used to control surface properties in a wide range. In this way protein adsorption, for instance, can be enhanced or largely prohibited at a pretreated surface.

Acknowledgement We would like to thank many groups leaders and graduate students, who are co-authors of the publications cited, for their motivated and skillful collaboration. Our work was funded by the Deutsche Forschungsgemeinschaft, the Fonds der Chemischen Industrie, and the Bundesministerium für Forschung und Technologie.

References

1. Möhwald H (1993) Rep Prog Phys 56:653
2. Evans DF, Wennerstrøm H (1994) The colloidal domain: where physics, chemistry, biology and technology meet. VCH, Weinheim
3. McConnell H, Desai RC (1991) Ann Rev Phys Chem 42:171; Knobler CM (1992) Ann Rev Phys Chem 43:207
4. Higgs PG, Joanny J-F (1991) J Chem Phys 94:1543
5. Cohen Stuart MA, Fleer GJ, Lyklema J, Norde W, Scheutjens JMHM (1991) Adv Colloid Interface Sci 34:477
6. Werner C, Eichhorn KJ, Grundke K, Simon F, Grählert W, Jacobasch H-J (1999) Colloid Surf A 156:3
7. Norde W (1986) Adv Colloid Interface Sci 25:267
8. Müller M, Werner C, Grundke K, Eichhorn KJ, Jacobasch H-J (1996) Macromol Symp 103:55
9. Netz RR, Joanny J-F (1998) Macromolecules 31:5123
10. Walter H, Müller-Buschbaum P, Gutmann JS, Lorenz-Haas C, Harrats C, Jérôme R, Stamm M (1999) Langmuir 15:6984

11. Mahltig B, Gohy J-F, Jérôme R, Bellmann C, Stamm M (2000) Colloid Polym Sci 278:502
12. Talingting MR, Ma Y, Simmons C, Webber SE (2000) Langmuir 16:862
13. Mahltig B, Gohy JF, Jérôme R, Stamm M (2001) J Polym Sci 39:709
14. Mahltig B, Müller-Buschbaum P, Wolkenhauer M, Wunnicke O, Wiegand S, Gohy J-F, Jérôme R, Stamm M (2001) J Colloid Interface Sci 242:36
15. Fromherz P (1971) Rev Sci Instr 46:1380; (1975) BBA 225:382
16. Engelking J, Ulbrich D, Menzel H (2000) Macromolecules 33:9026
17. Engelking J, Ulbrich D, Meyer WH, Schenk-Meuser K, Duschner H, Menzel H (1999) Mater Sci Eng C8–9:29
18. Helm CA, Laxhuber LA, Lösche M, Möhwald H (1986) Colloid Polym Sci 264:46
19. Ahrens H, Baltes H, Schmitt J, Möhwald H, Helm CA (2001) Macromolecules 34:4504
20. Ward AFH, Tordai L (1946) J Chem Phys 14:453
21. Motschmann H, Stamm M, Toprakcioglu C (1991) Macromolecules 24:3681
22. Sedlak M, Amis EJ (1992) J Chem Phys 96:826
23. Walter H, Harrats C, Müller-Buschbaum P, Jèrôme R, Stamm M (1999) Langmuir 15:1260
24. Kretzschmar G, Miller R (1991) Adv Colloid Interface Sci 36:65
25. Lucassen J, van den Tempel M (1972) Chem Eng Sci 27:1283
26. Avrami M (1939) J Chem Phys 7:1103; (1940) 8:212; (1941) 9:177
27. Engelking J, Menzel H (2001) Eur Phys J 5:87
28. Hönig D, Möbius D (1991) J Phys Chem 95:4590
29. Hénon S, Meunier J (1991) Rev Sci Instrum 62:936
30. Teppner R, Harke M, Motschmann H (1997) Rev Sci Instrum 68:4172
31. Lösche M, Sackmann E, Möhwald H (1983) Ber Bunsenges Phys Chem 87:848; Weiss RM, McConnell HM (1984) Nature 310:5972
32. Hatta E, Hosoi H, Ahyama H, Ishii T, MaKasa K (1998) Eur Phys J 2:347
33. Orrit M, Möbius D, Lehmann U, Maier H (1986) J Chem Phys 85:4966
34. Alexander S (1997) J Phys 38:983; Gennes PG (1980) Macromolecules 13:1069
35. Pincus P (1991) Macromolecules 24:2912
36. Fleer GJ, Cohen-Stuart MA, Schentjens JMHM, Cosgrove T, Vincent B (1993) Polymers at interfaces. Chapman and Hall, London
37. Simister EA, Lee EM, Thomas RK, Penfold J (1992) J Phys Chem 96:1373
38. Ahrens H, Förster S, Helm CA (1997) Macromolecules 30:8447
39. Ahrens H, Förster S, Helm CA (1998) Phys Rev Lett 81:4172
40. Pershan PS, Als-Nielsen J (1984) Phys Rev Lett 52:759; Als-Nielsen J (1986) Solid and liquid surfaces studied by synchrotron X-ray diffraction. In: Blanckenhagen W (ed) Structure and dynamics of surfaces. Springer, NY
41. Netz R, this volume
42. v Klitzing R, Espert A, Asnacios A, Hellweg T, Colin A, Langevin D (1999) Colloid Surf A 149:131; Theodoly O, Ober R, Williams CE (2001) Eur Phys J E 5:51
43. Borisov OV, Zhulina EB, Birshtein TM (1994) Macromolecules 27:4795; Zhulina EB, Birshein TM, Borisov OV (1995) Macromolecules 28:1491
44. Israels R, Leermakers FAM, Fleer GJ, Zhulina EB (1994) Macromolecules 27:3249
45. Manning GS (1969) J Chem Phys 51:934
46. Vaynberg KA, Wagner NJ, Ahrens H, Helm CA (1999) Langmuir 15:4686
47. Kjaer K, Als-Nielsen J, Helm CA, Laxhuber LA, Möhwald H (1987) Phys Rev Lett 58:2224
48. Als-Nielsen J, Kjaer K (1989) In: Rist T, Sherrington D (eds) Phase transitions in soft condensed matter, NATO ASJ series, vol 211. Plenum, NY, pp 13
49. de Meijere K, Brezesinski G, Möhwald H (1997) Macromolecules 30:2337
50. de Meijere K, Brezesinski G, Netz R, Möhwald H, to be published in Macromolecules
51. de Meijere K, Brezesinski G, Kjaer K, Möhwald H (1998) Langmuir 14:4204; de Meijere K, Brezesinski G, Pfohl T, Möhwald H (1999) J Phys Chem B42:8888
52. Symietz C, Brezesinski G, Möhwald H, to be published
53. Symietz C, Brezesinski G, Möhwald H, to be published

54. Kudaibergenov SE (1999) Adv Polym Sci 144:115
55. Mahltig B, Walter H, Harrats C, Müller-Buschbaum P, Jérôme R, Stamm M (1999) Phys Chem Chem Phys 1:3853
56. Walter H, Harrats C, Müller-Buschbaum P, Jérôme R, Stamm M (1999) Langmuir 15:1260
57. Potemkin II, Kramarenko EY, Khokhlov AR, Winkler RG, Reineker P, Eibeck P, Spatz JP, Möller M (1999) Langmuir 15:7290
58. Mahltig B, Jérôme R, Stamm M (2001) Phys Chem Chem Phys 3:4371
59. Mahltig B, Gohy J-F, Antoun S, Jérôme R, Stamm M (2002) Colloid Polym Sci 280:495
60. Mahltig B, Werner C, Müller M, Jérôme R, Stamm M (2001) J Biomater Sci Polym Ed 9:995
61. Mahltig B, Gohy J-F, Jérôme R, Buchhammer H-M, Stamm M (2002) J Polym Sci 40:338

Received November 2002

Adv Polym Sci (2004) 165:177–210
DOI 10.1007/b11270

Polyelectrolyte Membranes

Regine v. Klitzing[1] · Bernd Tieke[2]

[1] Stranski-Laboratorium, TU Berlin, Strasse des 17. Juni 112, 10623 Berlin, Germany
  E-mail: *klitzing@chem.tu-berlin.de*
[2] Institut für Physikalische Chemie, Universität zu Köln, Luxemburgerstrasse 116,
  50939 Köln, Germany

Abstract Two types of membrane are presented: free-standing films which are formed from aqueous polyelectrolyte solutions and membranes prepared by alternating electrostatic layer-by-layer assembly of cationic and anionic polyelectrolytes on porous supports. Layer-by-layer assemblies represent versatile membranes suitable for dehydration of organic solvents and ion separation in aqueous solution. The results show that the structuring of the polyelectrolytes in the liquid films and the permeability of the multilayer membranes depends on different internal and environmental parameters, for example molecular weight, polymer charge density, ionic strength, and temperature.

1
Introduction

Membranes play an important role in natural science and for many technical applications. Depending on their purpose, their shape can be very different. For instance, membranes include porous or non-porous films, either supported or non-supported, with two interfaces surrounded by a gas or by a liquid. Important properties of non-porous membranes are their *permeability* for certain compounds and their *stability*. In biological cells their main task is to stabilize the cell and to separate the cell plasma from the environment. Furthermore, different cells and cell compartments have to communicate with each other which requires selective permeability of the membranes. For industrial applications membranes are often used for separation of gases, liquids, or ions. Foams and emulsions for instance are macroscopic composite systems consisting of many membranes. They contain the continuous liquid phase surrounded by the dispersed gas phase (foams) or by another liquid (emulsions). Beside these application possibilities membranes give the opportunity to investigate many questions related to basic research, e.g. finite size effects.

This review focuses on two different types of polyelectrolyte membrane. The first part deals with thin free-standing liquid films formed from aqueous polyelectrolyte/surfactant solutions. Free-standing films are interesting in two respects. First, the film can be considered as the building block of a foam so that its properties affect the behavior of the whole macroscopic foam. In this context it is the molecular structure at and near the film surfaces rather than the structure of the film core which is important. Second, the free-standing film presents a slit pore which enables study of the effect of geometrical confinement on the structuring of polymers. The liquid free-standing film is highly relevant in this respect, since its thickness can easily be varied by changing the outer pressure.

The second section refers to polyelectrolyte membranes prepared by alternating electrostatic layer-by-layer assembly of cationic and anionic polyelectrolytes on porous supports. Mass transport across ultrathin polyelectrolyte multilayer membranes is described. The permeation of gas molecules, liquid mixtures, and ions in aqueous solution has been investigated. The studies indicate that the membranes are excellently suited for separation of alcohol/water mixtures under pervaporation conditions and for ion separation, e.g. under nanofiltration conditions.

The first part concentrates on *structure* while the second part focuses on the *permeability*. Both features are highly correlated and the influence on the membrane properties of electrostatics (e.g. ionic strength, polymer charge density, pH), temperature and the molecular weight of the polymers are studied on both types of membrane.

2
Aqueous Polyelectrolyte Membranes

Two large fields of colloidal science which have been investigated for several decades are associated with the studies presented in the first section–interactions between two opposing surfaces of free-standing aqueous films (so-called foam films) (e.g. Refs. [1, 2]) and the structuring of polyelectrolytes in aqueous solutions (e.g. Refs. [3–7]). With respect to general miniaturization the question arises about how this chain ordering is affected by a geometric confinement. Here *confinement* means that the dimensions of the total system are of the same order of magnitude as the correlation length of the polymer system (e.g. Debye length, correlation length, persistence length, radius of gyration). In the present study the polyelectrolytes are entrapped between the two interfaces of a horizontal single liquid free-standing surfactant (foam) film. A foam film gives the opportunity to adjust the thickness and therefore the confinement by changing the outer pressure. The film thickness is between 5 and 120 nm.

A quantitative measure of the sum of the interactions between the two film interfaces is the disjoining pressure Π. It is the excess pressure between the pressure in the film and the pressure of the corresponding liquid. This difference is due to interactions between the film interfaces which are mainly determined by electrostatic repulsion between (charged) interfaces, van der Waals attraction, steric repulsion between the adsorbed surfactant molecules, and structural forces. Disjoining pressure isotherms (disjoining pressure as a function of film thickness) have been measured in a so-called thin film pressure balance (TFPB).

2.1
Scientific Background

The electrostatic repulsion between two identically charged interfaces is described by an exponential decay:

$$\Pi_{el} = \Pi_0 \exp[-\kappa(h - 2h_0)] \tag{1}$$

where the Debye–Hückel screening length is $1/\kappa$ and the thickness of the adsorbed interfacial layer is h_0. Π_0 is connected to the surface potential Ψ_0 by the relation [8]:

$$\Pi(h) = 64kT\rho_\infty \gamma^2 \exp(-\kappa h) = \left(1.59 * 10^8\right)[c_{el}]\gamma^2 \exp(-\kappa h), [Pa] \tag{2}$$

where

$$\gamma = \tanh(ze\Psi_0/4kT) \tag{3}$$

In a simple foam film the thickness of the interface is similar to the length of a surfactant molecule. The thickness of the so-called common black film (CBF) is determined by the DLVO forces, and the thinner Newton black film (NBF) is stabilized by steric repulsion and does not contain any free solvent molecules. A transition from a CBF to a NBF can be induced by the addition of salt leading to a screening of the surface potential. This confirms the electrostatic nature of the repulsive force stabilizing the CBF. The transition from a CBF to a NBF corresponds to an oscillation of the disjoining pressure because of the attractive van-der-Waals forces. This attractive part of the isotherm is mechanically unstable, and it cannot be measured by a TFPB. But a step in film thickness from the thicker CBF to a thinner NBF is detected.

Beside the DLVO forces so-called structural forces can occur between two interfaces, i.e. in the film, due to the layering of molecules. This is related to an oscillatory concentration profile with a decay in amplitude from the interface towards the film bulk. The relation between the oscillatory concentration profile and the oscillatory pressure is still under discussion. An ostensive image of the oscillatory pressure is the layer-by-layer expulsion of the molecules which induces attractive depletion forces. With respect to the class of molecules this kind of force is also called solvation or hydration force. These structural forces have been observed for spherical molecules entrapped between two mica plates in a surface-force apparatus (SFA) (e.g. Ref. [8]). The period of the oscillation is connected to the diameter of a spherical molecule. Oscillatory forces also occur in free-standing films containing liquid crystals, colloidal particles (e.g. Ref. [1] and references therein) or micelles. Micellar systems have been investigated both in an SFA between solid interfaces [9] and in a TFPB between liquid interfaces [10, 11]. The step size scales with the surfactant concentration, c_s, as $c_s^{-1/3}$ [12]. This can be simply explained by a homogeneous distribution of "hard spheres" in three dimensions where the distance between two centers of mass is determined by the concentration of spheres. To the best of our knowledge, the stepwise thinning (i.e. stratification) of foam films containing micelles occurs if the surfactant is charged but it has yet been not published for non-ionic surfactants. This indicates that the electrostatic repulsion between the micelles could be responsible for film stratification.

2.2
Thin Film Pressure Balance (TFPB)

The measurements have been carried out in a thin film pressure balance using the porous-plate technique (Fig. 1). This method was developed by Mysels et al. [13] and subsequently enhanced by Exerova [14, 15] for measuring the disjoining pressure Π (h) in the film as a function of the film thickness h (disjoining pressure isotherm). The film is formed from an aqueous polymer solution over the (1 mm) hole drilled in a porous plate. The film holder is enclosed in a metal cell and the pressure inside the cell is changed using a syringe pump. The main contribution to the disjoining pressure within the film is the difference between the pressure inside (p_g) and outside (p_r) the cell. The film thickness is determined by an interferometric method [16]. In these circumstances the reflected intensity decreases with decreasing film thickness. This means that during the discontinuous transition from a thicker to a thinner film the "new" thinner film thickness is indicated as darker spots (see photograph in Fig. 1). The equilibrium thickness is reached when the intensity is constant over a period of 20 min. Further details are described elsewhere [17].

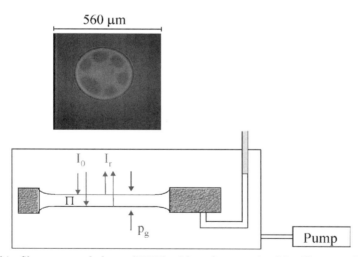

Fig. 1 Thin film pressure balance (TFPB) with a photograph of the film recorded by video microscopy

2.3
Materials

Polyelectrolytes of different molecular architecture are used.

2.3.1
Amphiphilic Diblock Copolymers

The amphiphilic diblock copolymer has been synthesized in the group of Stephan Förster (University of Hamburg). It consists of a poly(styrene sulfonic acid) block of 144 monomers and a hydrophobic poly(ethylethylene) block of 136 monomers. It is called PSSH–PEE in the following discussion. Synthetic details are described elsewhere [18]. The films of the diblock copolymers are made from pure aqueous polymer solutions without any further surfactant. The concentration of the diblock copolymers is given in g L^{-1}.

2.3.2
Branched Polyelectrolytes

The polycation PEI is a gift from BASF AG and it is available at three different degrees of branching with 100% secondary N atoms (linear), 50% and 38% secondary N atoms (depending on the number of bonds to the next carbon atoms one distinguishes between primary, secondary and tertiary nitrogen atoms). The 38%–N–PEI is irregularly branched and it is assumed to have an elliptic shape. The diameter is estimated from dynamic light-scattering measurements [19].

2.3.3
Linear Polyelectrolytes

For instance, a prototype of linear polyelectrolytes is a statistical copolymer consisting of positively charged diallyldimethylammonium chloride (DAD-MAC) monomers and neutral *N*-methyl-*N*-vinylacetamide (NMVA) monomers. It is a gift from Werner Jaeger from the Fraunhofer Institute for Applied Polymer Research in Golm/Potsdam (Germany). Details about the synthesis and the characterization are described in Ref. [20] for Poly(DADMAC) and in Ref. [21] for Poly(NMVA), and the copolymers. Unless stated otherwise the molecular weight is about 100,000 g mol^{-1}. The concentration of linear and branched polyelectrolytes is given in units of monomol L^{-1} which means the corresponding monomer concentration.

2.3.4
Surfactants

The nonionic surfactant $C_{12}G_2$ is used in the experiments at a fixed concentration of 10^{-5} mol L^{-1} ($0.1 \times$ cmc). The cationic hexadecyltrimethylammonium bromide (C_{16}TAB) was recrystallized four times from 10:1 ethyl acetate–ethyl alcohol and used at a fixed concentration of 10^{-4} mol L^{-1} ($0.1 \times$ cmc). Therefore, the surfactant concentration is always below the critical micellization concentration (cmc) and the critical aggregation concentration (cac) of the respective polymer/surfactant systems.

The free-standing films are either formed from aqueous surfactant/polyelectrolyte solutions or from pure aqueous diblock copolymer solutions.

2.4
Foam Films of Amphiphilic Diblock Copolymers

Amphiphilic diblock copolymers act as a surfactant and stabilize free-standing films. They are assumed to adsorb at the interface by analogy with low-molecular-weight surfactants: The hydrophobic part is collapsed at the interfaces and the hydrophilic part is directed towards the film core (Fig. 2a). Investigations of the structure at a *single* liquid interface (air/water) show that the amphiphilic diblock copolymers present polymer brushes which are anchored by the hydrophobic block at the interface [22, 23]. This structure is also assumed at the film surfaces. Fig. 3 shows the disjoining pressure iso-

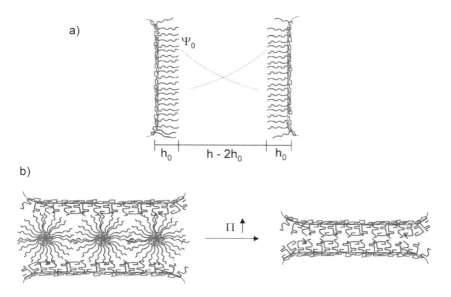

Fig. 2 (a) Schematic diagram of potential between the opposing polyelectrolyte brushes; (b) expulsion of one layer of micelles at higher diblock copolymer concentrations

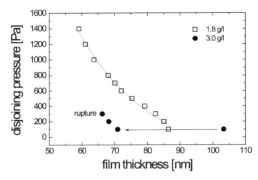

Fig. 3 Disjoining pressure isotherms of PSSH–PEE at two different concentrations—1.8 g L^{-1} and 3 g L^{-1}. The *solid line* corresponds to an exponential fit of the experimental data. For details see text (from Ref. [24])

therm at two different polymer concentrations. At a concentration of 1.8 g L^{-1} the film drains continuously and the data points can be fitted by Eq. (1). Under the assumption that the adsorbed layers have the same thickness of 22.5 nm as at the free interface, the fit results in a Debye length, κ^{-1}, of 15 nm and a surface potential, Ψ_0, of about 20 mV [24]. The exponential decay of the potential is sketched in Fig. 2a. The sign of the potential cannot be determined by the TFPB. The exponential decay is indicative of electrostatic repulsion between the two opposing polymer brushes (see Fig. 2a). The exponential decay is a hint that the two brushes do not interdigitate. Otherwise, the electrostatic repulsive force would increase with decreasing h as $1/h$ [25], instead of the found exponential increase.

At a higher concentration of 3 g L^{-1} the film shows discontinuous stratification. The isotherm shows a step in film thickness of approximately 32 nm. This step size is in good agreement with the size of a PSSH–PEE micelle in solution which has a diameter of approximately 30 nm [26]. It is assumed that by analogy with small surfactants a layer of micelles is entrapped between the two interfaces which is pressed out of the film (Fig. 2b). After the step, the film is thinner than for the lower concentration. The film is less stable, and it breaks at a pressure of about 300 Pa.

With increasing ionic strength the film becomes thinner, because of the electrostatic screening between the interfaces (Fig. 4). This is also observed for films of small surfactant molecules (see Scientific Background section) due to a decrease in the Debye length. Furthermore, the charged hydrophilic blocks coil with increasing ionic strength, which also contributes to the film thinning. This coiling could also explain the reduction in thickness and the stability at a higher polymer concentration, i.e. at a higher counterion concentration. Theoretical calculations by Pincus et al. [27] predict a scaling behavior of the film thickness, h, with the concentration of counterions as $h \propto c_{el}^{-1/3}$. In the present case an exponent of about −0.2 is found. The deviation could be explained by the finite volume of counterions which is not

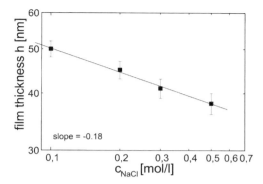

Fig. 4 Film thickness at a PSSH–PEE concentration of 3 g L^{-1} after the step in film thickness in dependence of the NaCl concentration. The *solid line* corresponds to a fit by a power law and results in an exponent of −0.2

taken into account in Pincus' theory. Vertical films of poly(*tert*-butylstyrene–PSS) (PtBS–PSS) which drain by gravitation forces showed the predicted exponent of −1/3 [28]. The reason for this difference in the experimental results is still under discussion.

2.5
Mixed Polyelectrolyte/Surfactant Films

In the following discussion the polyelectrolyte itself does not form any stable foam films, and the film has to be stabilized by additional surfactant. On the other hand, the surfactant itself does not form any stable films at such low concentrations below the cac as used in the presented experiments. This indicates synergistic effects between the two classes of compounds which lead to stable free-standing films. In the case of oppositely charged surfactants and polyelectrolytes surface-active complexes are assumed to be formed at the air/water interface below the bulk cac. This has been detected by a reduction in surface tension [29, 30], and by X-ray and neutron reflectometry [31, 32]. If one of the compounds is neutral or if both compounds have the same sign of charge addition of polyelectrolyte does not influence the surface tension, but nevertheless an increase in film stability is detected. This effect has not yet been clarified. Mixed polyelectrolyte/surfactant films also show stepwise thinning above a certain polyelectrolyte concentration. The steps are not reversible. By decreasing the pressure no step backward to the former branch of the isotherm is possible, and the thickness does not change significantly. Different combinations of surfactants and polyelectrolytes showed that the choice of the surfactant has an influence on the total film thickness but not on the step size of the disjoining pressure isotherms [17, 33, 34]. This leads to the conclusion that addition of polyelectrolytes induces the stratification of the films. The experimental data can be considered as parts of a damped oscillation as

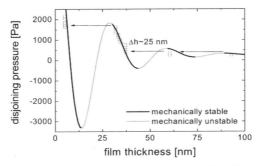

Fig. 5 Disjoining pressure isotherm of a $C_{12}G_2$/PDADMAC film. The solid line represents an exponentially damped cosine function. The mechanically stable and unstable parts are marked (from Ref. [33])

shown in Fig. 5. The attenuation explains the irreversibility of the steps: if the system is at a mechanically unstable point after increasing the pressure it jumps to the next mechanically stable state which is the next thinner isotherm branch. If, on the other hand, the pressure is decreased the system stays in a mechanically stable state and has no motivation to jump back to the former thicker isotherm branch. In relation to the stratification of films containing hard colloidal particles where the step size in film thickness is somehow related to the diameter of the particles, the question arises which characteristic length corresponds to the period of the oscillatory curve in the films containing polyelectrolytes. In the following discussion it is shown that the molecular architecture and the electrostatic interactions have an influence on the stratification behavior.

2.5.1
Branched Polyelectrolytes

Free-standing films which are formed from aqueous PEI/C_{16}TAB solution show a stepwise thinning. With increasing polymer concentration the step size decreases and the number of steps increases. Figure 6 shows the step size of a film containing branched PEI as a function of the corresponding concentration of monomer units for two different molecular weights of the PEI. The step size Δh scales with the polyelectrolyte concentration, c, as $c^{-1/3}$ [35]. By analogy with the spherical particles discussed above, a layer-by-layer expulsion of PEI molecules is assumed. The exponent $-1/3$ is explained by geometrical arguments–with increasing concentration of the branched polyelectrolytes in a three-dimensional system the distance between two molecules decreases in proportion to $c^{-1/3}$ in every dimension. At a fixed concentration of monomer units the number of PEI molecules decreases with increasing molecular weight. This leads to an increase in the intermolecular distance and therefore to an increase in the step size at the higher molecular weight. PEI is a weak polybase. At pH 9 (PEI in Milli-Q

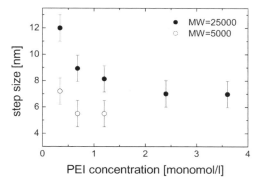

Fig. 6 Step size in the isotherms of films consisting of branched PEI and 10^{-5} mol L^{-1} $C_{16}TAB$ at a MW of PEI of 5000 and of 25,000 (from Ref. [35])

water) about 17% of the monomer units are charged (measured by streaming potential); at pH 4 (adjusted with HCl) about 74% of the units are charged. With decreasing polymer charge density the film stratification becomes less pronounced. The pressure needed to induce a step is lower and the film thickness between two steps can no longer be stabilized at pH 9 (Fig. 7). This leads to the conclusion that the layering of the PEI molecules is less ordered due to decreasing intermolecular electrostatic repulsion. This finding is also supported by drainage curves at these two pH values. The thinning velocity at pH 4 (i.e. higher polymer charge density) is much higher than that at pH 9. Due to electrostatic intermolecular repulsion the expulsion of the polyelectrolytes out of the film is much faster at pH 4 than at pH 9. Similar results are found for branched and linear polyelectrolytes. For linear PEI the disjoining pressure isotherm is less structured and the drainage process is slower than for branched ones. This could be due to a lower charge density in the case of the linear PEI. The number of charged monomers at a certain pH is almost independent of the degree of branching, but

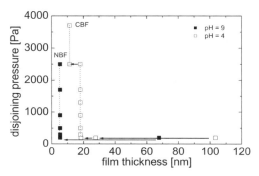

Fig. 7 Disjoining pressure isotherms of films formed from aqueous solutions containing 3.4×10^{-1} monomol L^{-1} PEI and 10^{-4} mol L^{-1} $C_{16}TAB$ at two different pH values

the volume of a branched PEI molecule is much smaller than that of a swollen coil of linear PEI. In addition, the linear chains interdigitate as described in the following section, which could again slow down the drainage process.

These results support the assumption that the oscillatory forces are caused by electrostatic repulsion between the layered polyelectrolytes. On the other hand the $c^{-1/3}$ dependency indicates that the step size is determined by the geometry but not by electrostatics. It is assumed that the concentration of free counterions around the PEI is quite low and that the step size Δh is:

$$\Delta h = d + 2\kappa^{-1} \tag{4}$$

where d is the diameter of the PEI molecules. The diameter of one PEI molecule of molecular weight 5000 is about 3 nm. The measured step size at a PEI monomer concentration of 0.34 monomol L^{-1} is about 7 nm. This corresponds to κ^{-1} of about 2 nm, which is indicative of an ionic strength of 0.025 mol L^{-1}. On the other hand, however, a PEI monomer concentration of 0.34 monomol L^{-1} (74% of the monomer units are charged at pH 4) corresponds to a concentration of monovalent counterions (i.e. ionic strength) of 0.25 mol L^{-1}. This difference in ionic strength means that only 10% of the counter ions are free and that the majority is entrapped within the branched molecules. The exponent of $-1/3$ can be explained by the fact that the ratio between free and entrapped counterions changes with the polymer concentration.

2.5.2
Linear Polyelectrolytes

Foam films containing linear polyelectrolytes also show stratification [17, 35–40]. In contrast to the branched polyelectrolytes the step size, Δh, scales with polyelectrolyte concentration as $\Delta h \propto c^{-1/2}$ which indicates another type of structuring within the film. This stratification behavior is independent of the chain stiffness and is observed for both flexible and semi-flexible chains [39] and even for worm-like micelles [41] which are regarded as "living" polymers. By analogy with the branched polyelectrolytes discussed above one could assume that the stratification is caused by a layering of polyelectrolyte coils and that the period is related to the radius of gyration. This would mean that the step size should change with the degree of polymerization. But the TFPB measurements show no effect of the chain length on the step size (Fig. 8), which means that a layering of polyelectrolyte coils cannot be the right model. Since only the polymers seem to induce the stratification the possible structuring of the corresponding aqueous polyelectrolyte solutions have been investigated by small-angle neutron scattering (SANS). The spectra show a broad peak which indicates an interaction between the chains [3–7, 33, 42, 43]. The peak position, q_{max}, shifts to higher q values with increasing polyelectrolyte concentration, c, and can be converted into a correlation length, ξ, by:

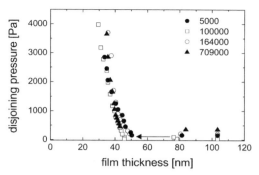

Fig. 8 Disjoining pressure isotherms of free-standing C_{16}TAB/PDADMAC films at different molecular weight of PDADMAC. The PDADMAC concentration is fixed at a corresponding monomer concentration of 5×10^{-3} monomol L^{-1}. Therefore the concentration of the polyelectrolyte molecules changes with the molecular weight. The C_{16}TAB concentration is 10^{-4} mol L^{-1} (from Ref. [33])

$$\xi = 2\pi/q_{max} \propto c^{-1/2} \tag{5}$$

on the assumption that the Bragg equation can be applied (Fig. 9). Such scaling behavior of the correlation length has been predicted by the isotropic model of de Gennes and several other groups [44–48] which assumes overlapping of the chain above a certain concentration $c*$ (semi-dilute regime). A kind of transient network is formed and the correlation length corresponds to the mesh size and the peak width in the SANS spectrum is related to the polydispersity of the mesh size. With increasing polyelectrolyte concentration the concentration of counterions increases. This leads to enhanced electrostatic screening between the chains which is related to a broadening of the correlation peak. Comparison of the correlation length obtained from SANS measurements with the step size in the disjoining pres-

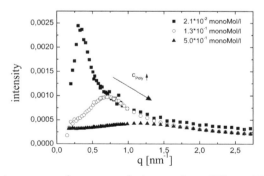

Fig. 9 SAN scattering curves of aqueous solutions at three different PDADMAC concentrations above the critical overlap concentration, $c*$, without additional salt. The polymer concentration is given in monomol L^{-1}; MW(PDADMAC)=100,000 g mol^{-1} (from Ref. [42])

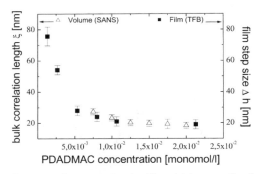

Fig. 10 Comparison between the step size in film thickness of a free-standing $C_{12}G_2/$ PDADMAC film (TFPB) and the correlation length of the respective aqueous bulk solution (SANS), calculated from the position of the structure peak in SANS spectra by Eq. (5), as a function of the polyelectrolyte concentration. The concentration of $C_{12}G_2$ was 10^{-5} mol L^{-1}. Adapted from Ref. [33]

sure isotherms shows that both lengths are similar (Fig. 10). This means that the interactions between the polyelectrolyte chains which induces a structuring of the chains in the aqueous solution are the same as the interactions in the film which lead to film stratification. On the assumption that the polyelectrolyte chains form a network-like structure in aqueous solution this structure can be obviously transferred to the film bulk. Hence, the period of the oscillatory disjoining pressure curve corresponds to the correlation length found in the bulk solution.

In comparison with the layered branched polyelectrolytes it is difficult to imagine the expulsion of linear polyelectrolyte chains which belong to a network. From fluorescence measurements at foam films containing dye-labeled polyelectrolytes it is known that the polyelectrolyte chains are pressed out of the film and that they are not collapsed within the film [40]. This means that as much of the corresponding monomers that belong to one mesh size are expelled during a transition from a thicker to a thinner film. But one polyelectrolyte chain is involved in several mesh sizes. The idea is that the network breaks down and is rebuilt much faster as the time resolution of the TFPB [33]. Up to a certain pressure the network rebuilds itself with a certain number of meshes in the film and above this pressure the network is rebuilt with one mesh less. The polyelectrolyte chains which are not involved in the network anymore are pressed out of the film. Using this image this would mean that the mesh size in solution and in the film is the same and that the film does not represent a confined geometry with respect to the mesh size. This could be explained by image charges induced by the air/liquid interface which extend the network to infinity. Furthermore, the surfactant at low concentrations has no effect on the step size which means that the interactions between polyelectrolytes and surfactant are negligible. This has been also observed by SANS measurements [33]. Oscillatory forces for linear polyelectrolytes have been also measured by Milling et al. with an AFM [49], and they also found that the correlation length scales with $c^{-1/2}$. To the best of our knowledge, no force oscillations for polymers have been observed in SFA studies. Oscillations in the concentration profile of a polymer system between two interfaces are predicted by theoretical models [50, 51], but until now no simulations exist of oscillatory forces of polyelectrolyte solution entrapped between two interfaces.

2.5.2.1
Effect of Electrostatics

Decreasing the charge density along the polyelectrolyte chain or increasing the concentration of low molecular mass salt leads to a reduction of the pressure which is necessary to induce a step in the film thickness. At high ionic strength or at a low degree of charge the stratification vanishes [37, 38]. The charge density has been changed by varying the ratio between cationic and neutral monomer units. Above a degree of charge of about 50% the steps occur one after the other at different pressures. Below a degree of charge of 50% all steps occur at the same low pressure. This leads to the conclusion that the electrostatic repulsion between the polyelectrolyte chains is reduced and that the structuring becomes "softer" at a low polymer charge density. In the example investigated, PDADMAC, above a degree of charge of 50% so-called charge renormalization takes place. The distance between two monomer units on the fully charged chain is almost half of the Bjerrum length, i.e. the distance between two ions where Coulombic interactions are compensated by the thermal energy. At a charge distance below the Bjerrum length counterion condensation [52–54] takes place which renormalizes the charge density. Therefore the effective charge density is the same above a degree of charge of 50% which seems to lead to similar disjoining pressure isotherms.

With increasing ionic strength the stratification of the film occurs at a lower pressure [38] which indicates that the structuring of the polyelectrolyte chains becomes "softer".

2.5.2.2
Effect of Temperature

With increasing temperature from 20 to 40 °C the film becomes less stable [55]. Decreasing stability with increasing temperature is also well-known for pure surfactant foam films and is caused by increasing fluctuations at the film interfaces. In surfactant/polyelectrolyte films temperature effects are even more pronounced. It is assumed that increasing motion of the polymer chains with increasing temperature further reduces film stability. Furthermore, the steps in film thickness are induced at a much lower pressure. It is assumed that chain motions at higher temperature also soften polyelectrolyte structuring within the film.

2.6
Summary and Conclusions

Disjoining pressure studies of free-standing liquid films containing polyelectrolytes show that the interactions between the interfaces are very sensitive to the molecular architecture of the polyelectrolytes. Stratification of the films occurs which is related to oscillatory structural forces. The size of the steps in film thickness scales in a different way with the polyelectrolyte concentration for different molecular architectures; this is due to different structuring of the polymers within the film. In the case of branched polyelectrolytes the step size scales as $\Delta h \propto c^{-1/3}$. This indicates expulsion of layers of the branched polymers by analogy with micelles entrapped within foam films above the cmc which results in the same scaling law. Linear polyelectrolytes lead to a scaling behavior as $\Delta h \propto c^{-1/2}$. It is assumed that a transient network of overlapping chains is formed within the film, as is assumed for polyelectrolyte solutions. The stratification of foam films made from amphiphilic diblock copolymers is assumed to be caused by expulsion of a layer of micelles previously entrapped within the film. A common feature of all types of these films is that the stratification and therefore the structuring of the polyelectrolytes is caused by electrostatic repulsion between the polyelectrolyte chains. It is reduced by decreasing the polymer charge density or increasing the ionic strength. The correlation length detected in semi-dilute solutions of linear polyelectrolytes is a statistical length, and it remains a general question how this length can be detected by force or disjoining pressure measurements.

3
Layer-by-Layer Assemblies of Polyelectrolytes as Separating Membranes

About a decade ago a new preparation route to organized molecular films was reported by Decher and coworkers [56–59]. The method is based on electrostatic layer-by-layer assembly of positively and negatively charged polyionic compounds on solid substrates. In a typical process a positively charged substrate is immersed in an aqueous solution of a negatively charged polyelectrolyte so that a monolayer of this polymer is adsorbed and the surface charge is reversed. After careful washing the substrate is now dipped into an aqueous solution of a positively charged polyelectrolyte, which also becomes adsorbed and adds a second polymer layer to the substrate. Multiple repetition of the adsorption steps leads to polyelectrolyte multilayers with precise control over thickness and polymer composition.

Soon after the first reports on the layer-by-layer adsorption appeared, the method was also used for surface modification of polymers [60, 61], and for the preparation of composite membranes [62–65]. Composite membranes were obtained by alternate dipping of porous supports into solutions of cationic and anionic polyelectrolytes so that an ultrathin separation layer was

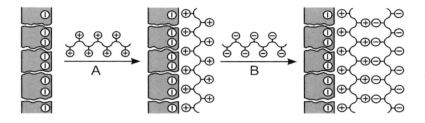

Support: PET-fleece (100 μm) + PAN-coating (80 μm, pore size 20 - 200 nm),

hydrophilised in O$_2$-plasma

Polyelectrolytes: A: PVA, PAH, P4VP, PEI, PDADMA, chitosan

B: PVS, PVSu, PSS, PAA, DEX

Fig. 11 Scheme of layer-by-layer assembly of polyelectrolytes on activated porous supporting membrane. The separation layer is obtained upon multiple repetition of steps A and B. In reality, pore diameters are 20 to 200 nm, polymer chains are less ordered and partially overlapping. Polyelectrolytes: *PVA*, poly(vinylamine); *PAH*, poly(allylamine hydrochloride); *PEI*; polyethyleneimine (branched), *P4VP*, poly(4-vinylpyridine); *PDADMA*, poly(diallyldimethylammonium chloride); *PVS*, poly(vinylsulfate);*PVSu*, poly(vinylsulfonate); *PSS*, poly(styrenesulfonate); *PAA*, polyacrylic acid; *DEX*, dextran sulfate (from Ref. [70])

formed, as outlined schematically in Fig. 11. In recent years, extensive studies have been performed on the separation capability of polyelectrolyte multilayer membranes. Gas separation [62–65], separation of liquid mixtures [65–73] under pervaporation [74] conditions, and ion separation [67, 75–78] have been studied. Research activities have recently been reviewed [79–81]. In the following discussion a survey of the most important results obtained by the author's research group is presented.

3.1
Formation of Polyelectrolyte Multilayer Membranes

In most of our studies [65–67, 69, 70, 72, 73, 75, 77, 78] porous PAN/PET supporting membranes were used consisting of a porous polyacrylonitrile (PAN) layer on a polyethylene terephthalate (PET) fleece. Before adsorption of the separation layer the membrane surface was rendered hydrophilic by treatment with oxygen plasma. Further data on membrane thickness, pore sizes and polyelectrolytes used for adsorption are supplied in Fig. 11. For preparation of the separation layer, the supporting membrane was alternately dipped into aqueous solutions of the cationic and anionic polyelectrolytes. Due to the repeated dipping, a thin homogeneous and dense skin layer consisting of the polyelectrolyte complex was formed on the substrate. After deposition of 60 layer pairs, the total thickness was in the range of a few hundred nanometers. The SEM picture in Fig. 12 shows that the polyelectrolytes

Fig. 12 Scanning electron micrograph of a composite membrane with polyelectrolyte multilayer (60 layer pairs) as skin layer and porous support consisting of PAN and PET layer (**a**) (pore sizes: 20 to 200 nm). In (**b**) the skin layer is shown at a larger magnification (from Ref. [90])

do not enter and block the pores, but the support is well coated with the polyelectrolyte skin layer. Since the support does not exhibit any transport selectivity by itself, all selectivities measured for the composite membrane can be exclusively ascribed to the polyelectrolyte separation layer. This enables study of the transport behavior of the polyelectrolyte complex layers with great precision.

3.2
Gas Permeation Studies

In order to make sure that a dense separation layer had formed, the permeation of gases through the membrane was investigated. First the dependence of the gas flow rate on the number of deposited polyelectrolyte layer pairs was studied [65–66]. For the permeation of argon across PAH/PSS separating membranes an inverse relation was found. The flow rate decreased from a value of 7.2 L cm^{-2} h^{-1} bar^{-1} for the bare support to values of 100 mL cm^{-2} h^{-1} bar^{-1} after a coating with 15 bilayers and to 8.7 mL cm^{-2} h^{-1} bar^{-1} for the sample coated with 60 bilayers, that is, the argon flow rate decreased by almost a factor of 10^3. Selectivities in gas flow were rather poor—60 P4VP/PSS bilayers only showed an ideal separation factor α (CO$_2$N2) of 1.51 [67], with PAH/PSS membranes no separation was found. Similar results were reported by other research groups [63]. The studies indicated that polyelectrolyte multilayer membranes are not well suited for gas separation.

3.3
Alcohol/water Pervaporation

The possibility of reducing the gas flow by a factor of 10^3 by layer-by-layer deposition of polyelectrolytes made it worthwhile to study liquid permeation and separation under pervaporation conditions. Earlier studies on solution cast, free-standing membranes of polyelectrolyte complexes reported by Paul and coworkers [82–84] had already shown a good selectivity in separation of alcohol/water mixtures under pervaporation conditions. Thus it was interesting to also study alcohol/water separation across the layer-by-layer assemblies of polyelectrolytes. Details of the home-made pervaporation apparatus can be found in the literature [79, 80]. First results obtained with PAH/PSS membranes 60 bilayers thick were not very promising, but it was found that flux and water content in the permeate were highly dependent on the preparation and operating conditions [65, 66]. Further investigations indicated that a suitable choice of polyelectrolyte was most important for achieving a good separation. The charge density of the polyelectrolytes, defined as the number of ionic groups per number of carbon atoms in the repeat unit of the polyelectrolyte complex, is especially important [69, 70]. Another important structural property is the acid/basic character of the ionic groups. Further parameters influencing ethanol/water separation are the number of adsorbed polyelectrolyte bilayers, the pervaporation temperature, the pH and ionic strength of the polyelectrolyte solutions used for membrane preparation, the molecular weight of the polymer and a thermal post-treatment of the membrane. In the following discussion the influence of the various parameters is briefly discussed.

3.3.1
Molecular Structure of the Polyelectrolytes

In order to describe the influence of polyelectrolyte structure on separation behavior it is useful to briefly discuss the structure of the separating membrane. In Fig. 13 [79, 80], a simplified structure model is shown. Although loop formation and interdigitation of chains from adjacent layers are neglected, as are incomplete ionization and neutralization, it clearly shows that the membrane exhibits a network structure with pores of defined size in the Ångstrom range, which are denoted as nanopores. Moreover, it can be derived from the model that cross-linking density and nanopore size will depend on the charge density of the polyelectrolyte complexes forming the membrane. Thus the charge density, ρ, of the polyelectrolyte complex was chosen as a structure parameter whose exact definition is given in Fig. 13. According to the structure model one can expect that membranes prepared from polyelectrolytes of high charge density consist of small, hydrophilic pores, which should be very permeable to water and less permeable for alcohols, whereas the use of polyelectrolytes of low charge density leads to membranes with larger, less hydrophilic pores. The larger pores can be expected to be more permeable to alcohol molecules and thus membranes prepared

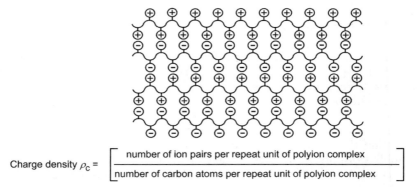

$$\text{Charge density } \rho_c = \left[\frac{\text{number of ion pairs per repeat unit of polyion complex}}{\text{number of carbon atoms per repeat unit of polyion complex}} \right]$$

Fig. 13 Simplified structure model of polyelectrolyte multilayer membrane and definition of charge density, ρ. The model does not take into account possible chain interdigitation and incomplete ionization (from Ref. [70])

from polyelectrolytes of low charge density should be less useful for alcohol/ water separation. The model was tested by studying ethanol/water separation across a variety of polyelectrolyte membranes with different ρ values ranging from 0.06 (for PDADMA/PSS) to 0.25 (for PVA/PVS, PEI/PVS and PVA/PVSu) [67, 69, 70, 72, 73]. As indicated in Fig. 14, the separation behavior of the various membranes is in good agreement with the predictions derived from the model of Fig. 13. With increasing value of ρ the total flux decreases, but the water content of the permeate strongly increases. This

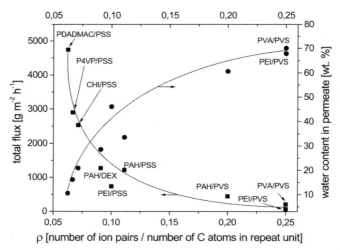

Fig. 14 Separation characteristics of polyelectrolyte multilayer membranes (60 layer pairs) as a function of the charge density, ρ. Feed solution: ethanol/water (93.8/6.2) (w/w); pervaporation temperature 58.5 °C (from Ref. [70]). Abbreviations are as in the caption to Fig. 11

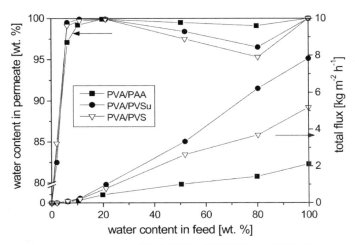

Fig. 15 Separation characteristics of PVA/PVSu, PVA/PVS and PVA/PAA membrane in ethanol/water pervaporation. Thickness of separation layer: 60 polyelectrolyte bilayers; pervaporation conditions: pH 6.8, no salt added to polyelectrolyte solution (from Ref. [73])

means that transport of water across the membrane is especially high, if the charge density of the polyelectrolyte complex is high so that the nanopores are very small and hydrophilic. The clarity of the correlation is surprising since the presence of any specific structural elements like phenyl rings, spacer groups or the nature of the ionic groups have not yet been taken into account. Therefore, in a further study the influence of the ionic groups on the separation characteristics was investigated [73]. For this purpose, ethanol/water pervaporation across three membranes of same high charge density containing PVA and different anionic polyelectrolytes with either sulfonate (PVSu), sulfate (PVS) or carboxylate (PAA) groups was investigated. As can be seen from Fig. 15, all membranes are able to separate the water quite efficiently. At low water content in the feed the membrane with the strongly acidic sulfonate groups is the best, because the sulfonate groups are most hydrophilic and even attract very small amounts of water from the mixture. However, at high water content in the feed, the strong hydrophilicity of PVSu is a drawback, because the polymer-bound ion pairs tend to hydrolyze, and the membrane begins to swell. As a consequence, the flux goes up, and the separation decreases. If the water content in the feed exceeds 20 wt%, the PVA/PAA membrane is better suited for separation, but the comparatively poor hydrophilicity of the carboxylate groups results in the flux across this membrane always being rather low.

3.3.2
Number of Deposited Layer Pairs

From the successive adsorption of polyelectrolyte layers on a porous support one can, first, expect that the pores of the supporting membrane become coated and, second, that the thickness of the separating membrane gradually increases. Consequently, the total flux should decrease and eventually–when all the pores are coated–the separation should improve. Experiments indicated that this is indeed the case [66, 67]. With increasing number of deposited PVA/PVS bilayers, the water content of the permeate increases from 10 wt% for 10 bilayers to 92 wt% for 50, and 98 wt% for 60 bilayers, if the ethanol/water feed solution contains 6.3 wt% water. Simultaneously the total flux decreases from about 1 kg m^{-2} h^{-1} to about 150 g m^{-2} h^{-1} for the 50 bilayer sample, and to 110 g m^{-2} h^{-1} for 60 bilayers. The decrease mainly reflects the usual inverse relationship between flux and thickness of a membrane. From the measurements it can be derived that the adsorption of at least 50 polyelectrolyte bilayers on the highly porous PAN/PET support is necessary to get an optimum separation. Most of our experiments were therefore carried out on membranes containing 60 bilayers.

3.3.3
Pervaporation Temperature

Increasing the pervaporation temperature is quite often useful in order to improve the flux and separation efficiency of a membrane. This also holds for polyelectrolyte multilayer membranes [66]. For 2-propanol/water pervaporation across PVA/PVS membranes it was found, for example, that the total flux increases by a factor of ten upon changing the pervaporation temperature from 25 to 60 °C, whereas the water content of the permeate also increased slightly. The reason is that increasing the temperature mainly increases the flux of water across the membrane, which is favorable for the separation characteristics. For an optimum separation all pervaporation measurements were thus carried out at about 60 °C, the highest temperature technically available with our apparatus.

3.3.4
pH and Ionic Strength of Polyelectrolyte Solutions Used for Membrane Preparation

The conformation of a polyelectrolyte in solution is known to be highly dependent on the pH and ionic strength of the solution. Previous studies have shown that these also affect ionization of the polar groups and chain conformation of the adsorbed polymer so that thickness, surface roughness and charge density of the polyelectrolyte multilayers become a function of pH and ionic strength of the solutions used for layer-by-layer assembly [85–89]. In fact, SEM pictures of PAH/PSS membranes prepared either from acid or neutral solution, or acid solution in presence of sodium chloride, indicate

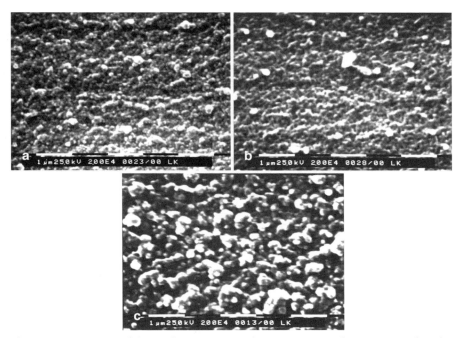

Fig. 16 SEM pictures of the surface structure of a PAH/PSS membrane prepared under different conditions of the polyelectrolyte solution: (**a**) neutral pH, no salt added; (**b**) pH 1.7, no salt added; (**c**) pH 1.7; 0.1 mol L^{-1} NaCl (from Ref. [90])

large differences between the surface structures (Fig. 16) [90]. Thus it is likely that the separation behavior is also affected. In order to study the influence on the separation characteristics, two PVA/PVS membranes were prepared from either neutral or acid solution at pH 6.8 and 1.7, respectively. A third membrane was prepared from polyelectrolyte solutions at pH 1.7 containing additional sodium chloride (concentration 0.1 mol L^{-1}) [73]. As shown in Fig. 17, the poorest separation was found for the membrane prepared from aqueous solution at pH 1.7 without additional sodium chloride. The presence of salt, or the higher pH of 6.8 positively influenced the separation capability, while the flux was affected only if salt was present in the polyelectrolyte solution. In this case a substantially lower total flux was observed. This might be because the presence of salt in the polyelectrolyte solution reduces the size of the polyelectrolyte coils. Unfolding of the coils during adsorption is rendered more difficult so that the polymers are partially adsorbed as coils. Consequently the thickness of the layers increases, the flux is lowered, and the separation is improved. If the membrane is prepared at pH 6.8, the polar groups of the polyelectrolytes are present in a completely ionized state, a very regular network structure of high density is formed and the thickness of the membrane is low. Consequently, good separation and a high flux are obtained.

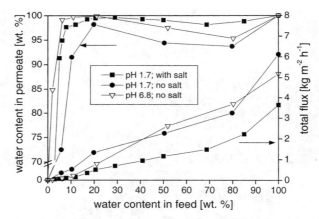

Fig. 17 Separation characteristics of PVA/PVS membranes (60 bilayers) prepared under different conditions of pH and NaCl content of the polyelectrolyte solution. Pervaporation temperature: 58.5 °C (from Refs. [73, 81])

3.3.5
Molecular Weight of Polyelectrolytes

Another important parameter influencing the separation characteristics is the molecular weight of the polyelectrolytes. The effect of molecular weight was studied on PVA/PAA membranes with PAA(l) of low molecular weight of 5000 and PAA(h) of high molecular weight of 250,000 [73]. While the water enrichment was the same for the two membranes, the flux of the membrane containing PAA(l) was only a quarter of the value found for the other. It was assumed [73] that the short polymer chains are able to enter and stop the pores of the supporting membranes so that the flux is decreased. Moreover, the use of PAA(l) also reduces the long-term stability of the membrane, probably because the polymer chains are so short that they are gradually leached out during the pervaporation experiments.

3.3.6
Annealing of the Membrane

For many of the membranes, annealing subsequent to adsorption of the separation layer was advantageous in order to improve the separation characteristics. Upon annealing of PEI/PVS and PAH/PSS membranes at 90 °C, for example, the separation of ethanol/water (9:1) mixtures could be highly improved [69]–for PEI/PVS the water content of the permeate increased from 67.8 to 96.7 wt%, while the flux decreased from 70 to 45 g m^{-2} h^{-1}, and for PAH/PSS, the water content of the permeate increased from 28.5 to 80.9 wt%, while the flux decreased from 1270 to 240 g m^{-2} h^{-1}. However, membranes containing PVA as the cationic polyelectrolyte could not be improved in their separation behavior upon annealing. SFM pictures of a PAH/

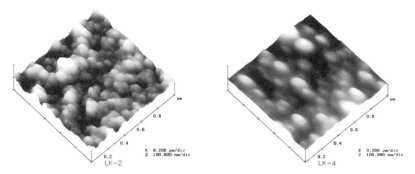

Fig. 18 Scanning force microscopic pictures (tapping mode) of the surface structure of a 60 bilayer PAH/PSS membrane prior (*left*) and subsequent to annealing at 90 °C (*right*) (from Ref. [90])

PSS membrane in Fig. 18 indicate changes in the surface structure upon annealing [90]. The height differences are reduced from 22.9 to 9.9 nm, the membrane appears much smoother and more homogeneous, which goes along with the improvement of the separation behavior.

3.3.7
Separation of Various Alcohol/Water Mixtures

The good results in ethanol/water separation obtained with the PVA/PVS and PVA/PVSu membranes suggested extending the studies to other alcohol/water mixtures. In Fig. 19a–d, the separation behavior of various alcohol/water mixtures using PVA/PVSu membranes is compiled [73]. It turned out that flux and separation increased if the hydrophilicity of the alcohols decreased. For a t-butanol/water (9:1) mixture, for example, a flux, J, of 2 kg m^{-2} h^{-1}, a water content in the permeate of 99.99%, a separation factor α of 8.500, and a separation efficiency αJ (also denoted as 'pervaporation separation index') of 1.95×10^4 were found, if a PVA/PVSu membrane was used and the pervaporation was carried out at 58.5 °C. The PVA/PVS membrane showed very similar separation behavior [72]. There are two reasons for the excellent separation of the more hydrophobic alcohols from water. First, the strength of the hydrogen bonds between alcohol and water molecules decreases with increasing number of carbon atoms in the alcohol, making the separation easier. Second, the solubility of the alcohol molecules in the hydrophilic membrane decreases with increasing alkyl chain length rendering transport across the membrane more difficult.

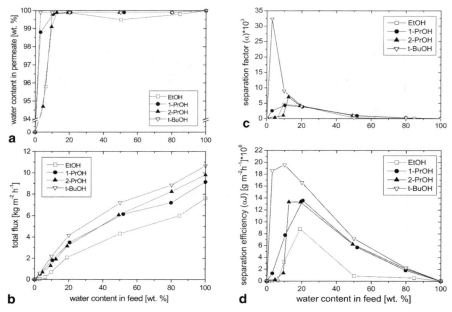

Fig. 19 Separation characteristics of a PVA/PVSu membrane (60 bilayers) for various alcohol/water feed mixtures. Plot of water content in permeate (**a**), total flux (**b**), separation factor α (**c**) and separation efficiency (**d**) versus water content in feed mixture. Pervaporation temperature 58.5 °C (from Ref. [73])

3.4
Transport of Ions

From earlier studies [91, 92] it is known that the bilayer casting of two aqueous polyelectrolyte solutions of opposite charge leads to so-called bipolar membranes. These membranes are able to separate mono- and divalent ions, because the divalent ions receive much stronger repulsive forces from the equally charged layers and attractive forces from the oppositely charged layers than the monovalent ions. This is true for both anions and cations so bipolar membranes represent barriers for ions of high charge density in general.

3.4.1
Rejection Model of Multi-Bipolar Polyelectrolyte Membrane

The polyelectrolyte multilayers consist of an alternating sequence of molecular layers of cationic and anionic polyelectrolytes and can therefore be regarded at as a multi-bipolar membrane on a molecular level, which likewise should be useful for ion separation. A corresponding model for the ion rejection of the multibipolar membrane [75] is shown in Fig. 20. The model

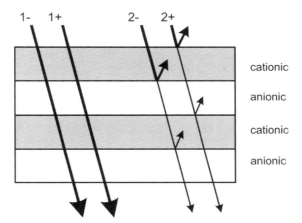

Fig. 20 Rejection model of the multi-bipolar polyelectrolyte multilayer membrane (from Ref. [76])

implies separation of monovalent and divalent ions, which becomes progressively more effective if the number of deposited polyelectrolyte layers is increased and which is independent of the surface charge of the membrane.

3.4.2
Ion Permeation at Ambient Pressure

In a number of studies [75–77] ion transport across the polyelectrolyte multilayer membranes was investigated. Most studies were concerned with ion permeation along a concentration gradient at atmospheric pressure. Experimental details are given elsewhere [75, 77]. From the permeation measurements, preferential transport of monovalent ions across the membranes was evident. Using PAH/PSS separating membranes, theoretical separation factors α (Na^+/Mg^{2+}) up to 112.5 and α (Cl^-/SO_4^{2-}) up to 45.0 could be measured [75]. The ion permeation was controlled by factors such as the number of deposited layers, the charge density of the polyelectrolytes, and the pH and ionic strength of the polyelectrolyte solutions used for membrane preparation.

If the number of deposited layers was increased, the permeation rates, P_R, of all the salts decreased, but for salts with divalent or trivalent metal ions the decrease was stronger than for salts with monovalent ions (Fig. 21). Consequently the selectivity increased if a greater number of layers was adsorbed, as predicted by the model discussed above. However, different from pervaporation separation, for example, a good separation was already found for separation layers consisting of five polyelectrolyte bilayers only (α (Na^+/Mg^{2+})=51) [76]. The ion separation also increased if polyelectrolytes of progressively higher charge density were used for preparation of the membrane. With increasing charge density the cross-linking density of the network in-

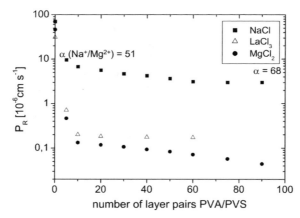

Fig. 21 Plot of permeation rates P_R of NaCl, MgCl$_2$ and LaCl$_3$ versus the number of ad-sorbed PVA/PVS bilayers. Separating membranes were prepared from polyelectrolyte solutions containing 0.1 mol L^{-1} NaCl, pH 1.7 (from Ref. [76])

creased and the P_R values of all ions decreased. However, again the decrease was stronger for the divalent than for the monovalent ions so that the selectivity in ion transport increased.

3.4.3
The Influence of the Charge Density of the Permeating Ions

In order to gain information on the transport mechanism the permeation rates of a number of alkali and alkaline earth metal chlorides, copper, zinc, and lanthanum chloride, and sodium and magnesium sulfate were determined. In an additional experiment, the concentration of the permeating salt solution was varied from 1 mmol L^{-1} to 3 mol L^{-1}. All measurements were carried out on membranes consisting of 60 PVA/PVS bilayers [76].

The permeation rates of the metal chlorides were found [76] to be inversely proportional to the charge density of the metal ions. The reason is that ions of low charge density only display weak interactions with the polyelectrolyte membrane and thus are able to permeate rapidly across the membrane whereas ions of high charge density and strong interactions only exhibit low P_R values. The influence of the charge density on P_R is even recognizable for the series of alkali and alkaline earth metal chlorides: P_R values drop in the sequence K$^+$>Na$^+$>Li$^+$ and Ba^{2+}>Ca^{2+}>Mg^{2+}, that is in the direction of increasing charge density of the non-hydrated cations (Fig. 22). The same is true for the anions so that the P_R values of electrolytes generally decrease in the sequence 1,1-electrolyte>2,1-electrolyte$\geq$1,2-electrolyte>2,2-electrolyte. For example, the ratio of the P_R values of sodium and magnesium chloride and sodium and magnesium sulfate is 3000:10:30:1 [76]. One may find the correlation between P_R and the non-hydrated cation radii in Fig. 22 surprising, because ions in aqueous medium are generally believed

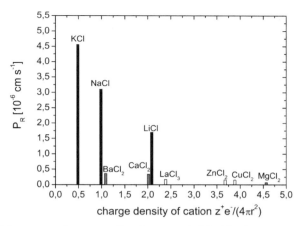

Fig. 22 Plot of the permeation rates of various metal chlorides in aqueous solution (feed concentration: 0.1 mol L^{-1}) against the charge density $z^+e^-/4\pi r^2$ of the non-hydrated metal cations (z^+: charge number of ion; r: radius of metal cation)

to be strongly hydrated and the transport behavior is rather determined by the hydrodynamic radius, which decreases from lithium to potassium ions, for example. However, the results clearly indicate that in the membrane hydration only plays a minor role. Instead interactions between the water molecules and charged segments of the polyelectrolyte chains are stronger and the available energy is not sufficient to cause hydration of the permeating ions.

3.4.4
Permanent Incorporation of Ions

From Fig. 22, very low permeation rates are evident for barium, calcium, copper and lanthanum chloride. In a recent study [78] it was shown that the interactions of certain divalent and multivalent ions as for example Cu^{2+}, La^{3+}, Ba^{2+} and $[Fe(CN)_6]^{4-}$ with the polyelectrolyte chains are so strong that these ions become fixed in the membrane. In most cases the permanent incorporation of the ions is accompanied by an increase of the network density of the membrane. As a consequence, the membrane becomes less permeable for all ions. This aging process was quantitatively investigated by studying multiple alternating permeation of aqueous NaCl and $CuCl_2$ solutions across a PVA/PVS membrane [78]. Although the membrane was carefully washed after each salt permeation, copper ions were permanently incorporated and–after fivefold alternating permeation of the two salt solutions–the P_R values of both salts dropped to half of the original values (Fig. 23). If the same experiment was carried out with KCl and $K_4Fe(CN)_6$, the decrease was as high as 70%. Only $BaCl_2$ was an exception, as also indicated in Fig. 23. Because of its high affinity for the sulfate ions of PVS it is able to replace the

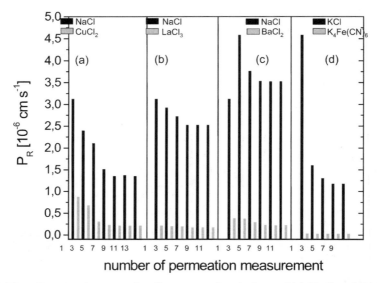

Fig. 23 Plot of permeation rates P_R of aqueous salt solutions of (**a**) NaCl and CuCl$_2$, (**b**) NaCl and LaCl$_3$, (**c**) NaCl and BaCl$_2$, and (**d**) KCl and K$_4$Fe(CN)$_6$, determined for multiple alternating permeation of the respective salt solutions across a PVA/PVS membrane (60 bilayers). The P_R values gradually change upon repeated permeation due to permanent incorporation of the divalent, trivalent and tetravalent ions in the membrane (from Ref. [78])

ammonium ions of PVA from the polyelectrolyte complex. As a consequence, the network density of the membrane is lowered and the permeation rates are increased.

3.4.5
The Influence of Electrolyte Concentration

In a further set of experiments, the influence of the concentration of the permeating salt solution on P_R values and selectivities was investigated [76]. It was found that the permeation rates gradually increased with the salt concentration. The effect was weak in the concentration range from 1 mmol L^{-1} to 0.1 mol L^{-1}, and became very strong at concentrations above 0.3 mol L^{-1}. Simultaneously, the selectivity decreased. While for 1 millimolar electrolyte solutions the theoretical separation factor α (NaCl/MgCl$_2$) was 60, it was only 10 at concentrations of 0.3 mol L^{-1}. This indicates a structural change of the membrane, the higher electrolyte concentration causing the formation of small pores and channels in the membrane through which the electrolyte solution can flow. The effect is partially reversible and can be explained by considering the equilibrium:

$$P^+X^- + P^-M^+ \rightleftarrows P^+P^- + M^+X^- \tag{6}$$

of the formation of the polyelectrolyte complex P^+P^- constituting the membrane. According to Eq. (6), the presence of locally high concentrations of electrolyte M^+X^- causes the equilibrium to be shifted to the left, and thus the polymer-bound ion pairs P^+P^- dissociate and the network begins to open.

3.4.6
Ion Transport under Nanofiltration and Reverse Osmosis Conditions

Very recent studies [93] have investigated the pressure-driven transport of inorganic ions across the polyelectrolyte multilayers. Using a home-made dead-end apparatus, ion permeation was investigated under nanofiltration and reverse osmosis conditions, that is along a concentration and pressure gradient across the membrane. The transport of aqueous sodium and magnesium chloride solutions with concentrations of 1 and 10 mmol L^{-1} was investigated using a PVA/PVS multilayer membrane 60 bilayers thick. The permeation flux of all solutions increased linearly with pressure, the slope being practically equal to the pure water flux. At operating pressures from 5 to 20 bar sodium ion rejection increased from 84 to 92% and at an operating pressure of 40 bar, a rejection of 93.5% was reached. For the two sodium chloride solutions of different concentration the same rejection was found within the experimental error. For the magnesium chloride solutions ion rejection was complete over the whole pressure range. The total rejection can be ascribed to strong electrostatic interactions of the permeating magnesium ions with the charged groups of the membrane matrix. The study [93] indicates that the polyelectrolyte multilayers show the typical behavior of nanofiltration membranes with partial rejection of monovalent and complete rejection of divalent ions at moderate applied pressures. In this pressure range the membranes are suitable for water softening. At high operating pressure nearly complete ion rejection is found so that the membranes are even suitable for water desalination under reverse osmosis conditions.

3.5
Summary and Conclusions

These investigations demonstrated that alternating electrostatic layer-by-layer adsorption of cationic and anionic polyelectrolytes is a useful method of preparing ultrathin, dense separating membranes on porous supports. The thickness of the separating membranes is in the nanometer range and can be adjusted with monolayer precision. The separating membranes consist of a nanoporous, hydrophilic network structure of the polyelectrolyte complexes, which is easily permeable to small hydrophilic molecules and ions of low charge density, such as potassium, sodium, and chloride ions. Larger, less hydrophilic alcohols, as for example propanol or *t*-butanol, and ions of high charge density such as divalent metal cations or multivalent metal complex ions are largely rejected. Due to the highly selective mass transport, the polyelectrolyte membranes are suitable for dehydration of organic solvents, especially alcohol/water separation, and for separation of monovalent and multivalent ions, e.g. under nanofiltration conditions, and for water desalination under reverse osmosis conditions.

The selectivity in the mass transport behavior is probably one of the most interesting properties of polyelectrolyte multilayer systems in general. The suitability of the multilayer systems as ion separation membranes has been confirmed by other research groups [76] and attempts to further improve ion selectivity have been reported [94–96], e.g. by chemical cross-linking of the membranes [94]. In other studies attempts have been made to understand the separation behavior on a theoretical [97] and experimental basis [98, 99], and to improve the performance of the membranes in alcohol/water pervaporation [71]. High permeability and selective mass transport render the polyelectrolyte multilayer membranes interesting not only as separating membranes but also for applications in the sensors. The main aspects of the transport behavior obtained from the studies using flat membranes also aid understanding of the transport behavior of polyelectrolyte hollow capsules [100], which might be useful as drug carriers in medicine [101]. Future work will be concerned with the use of the polyelectrolyte multilayers in electrodialysis and in ultrafiltration and microfiltration. Moreover, it appears interesting to incorporate permselective components in the membranes to improve the separation capability. The use of chiral compounds may even enable separation of enantiomers.

Acknowledgement We thank the DFG (Schwerpunkt 1009 "Polyelektrolyte mit definierter Molekülarchitektur") for financial support.

References

1. Ivanov IB, Dimitrov DS (1988) In: Ivanov IB (ed) Thin liquid films. Marcel Dekker, New York, p 379
2. Exerowa D, Kruglyakov K (1998) Foam and foam films. Elsevier, Amsterdam
3. Nierlich M, Williams C, Boué F, Cotton JP, Daoud M, Farnoux B, Jannink G, Picot C, Moan M, Wolff C, Rinaudo M, de Gennes PG (1979) J Phys 40:701
4. Förster S, Schmidt M (1995) Adv Polym Sci 120:51
5. Nishida K, Kaji K, Kanaya T (1995) Macromolecules 28:2472
6. Essafi W, Lafuma F, Williams CE (1999) Eur Phys J B 9:261
7. Claesson PM, Bergström M, Dedinaite A, Kjellin M, Legrand JF, Grillo I (2000) J Phys Chem B 104:11689
8. Israelachvili J (1998) In: Israelachvili J (ed) Intermolecular and surface forces. Academic Press, New York, p 266
9. Richetti P, Kekicheff P (1992) Phys Rev Lett 68:1951
10. Bergeron V, Radke CJ (1992) Langmuir 8:3020
11. Sonin AA, Langevin D (1993) Europhys Lett 22:271
12. Bergeron V, Claesson PM (2002) Adv Colloid Interface Sci 96:1
13. Mysels KJ, Jones MN (1966) Discuss Faraday Soc 42:42
14. Exerova D, Schedulko A (1971) Chim Phys 24:47
15. Exerova D, Kolarov T, Khristov KHR (1987) Colloid Surf A 22:171
16. Schedulko A (1967) Adv Colloid Interface Sci 1:391

17. v Klitzing R, Espert A, Asnacios A, Hellweg T, Colin A, Langevin D (1999) Colloid Surf A 149:131
18. Förster S, Hermsdorf S, Leube W, Schnablegger H, Regenbrecht M, Akari S (1999) J Phys Chem B103:6657
19. Hellweg T, Henry-Toulme N, Chambon M, Roux D (2000) Colloid Surf A 163:71
20. Dautzenberg H, Görnitz E, Jaeger W (1998) Macromol Chem Phys 199:1561
21. Ruppelt D, Kötz J, Jaeger W, Friberg SE, Mackay RA (1997) Langmuir 13:3316
22. Ahrens H, Förster S, Helm CA (1997) Macromolecules 30:8447
23. Ahrens H, Förster S, Helm CA (1998) Phys Rev Lett 81:4172
24. Kolaric B, Förster S, v Klitzing R (2001) Progr Colloid Polym Sci 117:195
25. Ohshima H (1999) Colloid Polym Sci 277:535
26. Regenbrecht M, Akari S, Förster S, Möhwald H (1999) J Phys Chem B 103:6669
27. Pincus P (1991) Macromolecules 24:2912
28. Guenoun P, Scalchli A, Sentenac D, Mays JW, Benattar JJ (1995) Phys Rev Lett 74:3628
29. Asnacios A, Langevin D, Argillier JF (1996) Macromolecules 29:7412
30. Asnacios A, v Klitzing R, Langevin D (2000) Colloids Surf A 167:189
31. Stubenrauch C, Albouy PA, v Klitzing R, Langevin D (2000) Langmuir 16:3206
32. Taylor DJF, Thomas RK, Penfold J (2002) Langmuir 18:4748
33. v Klitzing R, Kolaric B, Jaeger W, Brandt A (2002) Phys Chem Chem Phys 4:1907
34. Kolaric B, Jaeger W, Hedicke G, v. Klitzing R, J Phys Chem B, pusblished as ASAP article since June 2003
35. v Klitzing R, Kolaric B (2003) Progr Colloid Polym Sci 122:122–129
36. Bergeron V, Langevin D, Asnacios A (1996) Langmuir 12:1550
37. Asnacios A, Espert A, Colin A, Langevin D (1997) Phys Rev Lett 78:4974
38. Kolaric B, Jaeger W, v Klitzing R (2000) J Phys Chem B 104:5096
39. v Klitzing R, Espert A, Colin A, Langevin D (2001) Colloids Surf A 176:109
40. Toca-Herrera JL, v Klitzing R (2002) Macromolecules 35:2861
41. Espert A, v Klitzing R, Poulin P, Colin A, Zana R, Langevin D (1998) Langmuir 14:4251
42. v Klitzing R (2000) Tenside Surf Det 37:338
43. Theodoly O, Tan JS, Ober R, Williams CE, Bergeron V (2001) Langmuir 17:4910
44. de Gennes PG, Pincus P, Velasco RM, Brochard F (1976) J Phys France 37:37
45. Odijk T (1977) J Polym Sci Polym Phys Ed 15:688
46. Skolnick J, Fixman M (1977) Macromolecules 10:944
47. Dobrynin AV, Colby RV, Rubinstein M (1995) Macromolecules 28:1859
48. Barrat JL, Joanny JF (1996) Adv Chem Phys 94:1
49. Milling AJ (1996) J Phys Chem 100:8986
50. Chatelier X, Joanny JF (1996) J Phys II France 6:1669
51. Yethiraj A (1999) J Chem Phys 111:1797
52. Oosawa F (1957) J Polym Sci 23:421
53. Manning GS (1969) J Chem Phys 51:924
54. Ray J, Manning GS (1997) Macromolecules 30:5739
55. Tian G, v Klitzing R, unpublished results
56. Decher G, Hong JD (1991) Makromol Chem Macromol Symp 46:321
57. Decher G, Hong JD (1991) Ber Bunsenges Phys Chem 95:1431
58. Decher G, Hong JD, Schmitt J (1992) Thin Solid Films 210/211:831
59. Decher G (1997) Science 277:1232
60. Chen W, McCarthy TJ (1997) Macromolecules 30:78
61. Delcorte A, Bertrand P, Wischerhoff E, Laschewsky A (1997) Langmuir 13:5125
62. Stroeve P, Vasquez V, Coelho MAN, Rabolt JF (1996) Thin Solid Films 284/285:708
63. Leväsalmi JM, McCarthy TJ (1997) Macromolecules 30:1752
64. Zhou P, Samuelson L, Shridara Alva K, Chen CC, Blumstein RB, Blumstein A (1997) Macromolecules 30:1577
65. van Ackern F, Krasemann L, Tieke B (1998) Thin Solid Films 329:762
66. Krasemann L, Tieke B (1998) J Membr Sci 150:23
67. Krasemann L, Tieke B (1999) Mater Sci Eng C 8/9:513

68. Meier-Haack J, Rieser T, Lenk W, Lehmann D, Berwald S, Schwarz S (1999) Chem Ing Tech 71:839
69. Krasemann L, Tieke B (2000) Chem Eng Technol 23:211
70. Krasemann L, Toutianoush A, Tieke B (2001) J Membr Sci 181:221
71. Meier-Haack J, Lenk W, Lehmann D, Lunkwitz K (2001) J Membr Sci 184:233
72. Toutianoush A, Krasemann L, Tieke B (2002) Colloids Surf A 198–200:881
73. Toutianoush A, Tieke B (2002) Mater Sci Eng C 22:459
74. Neel J (1991) In: Huang RYM (ed) Pervaporation membrane separation processes. Elsevier, Amsterdam, p 42
75. Krasemann L, Tieke B (2000) Langmuir 16:287
76. Harris JJ, Stair JL, Bruening ML (2000) Chem Mater 12:1941
77. Toutianoush A, Tieke B (2001) In: Moebius D, Miller R (eds) Novel methods to study interfacial layers, Studies in Surface Science Series. Elsevier, Amsterdam, p 416
78. Toutianoush A, Tieke B (2003) Mater Sci Eng C 22:135
79. Tieke B, Krasemann L, Toutianoush A (2001) Macromol Symp 163:97
80. Tieke B, van Ackern F, Krasemann L, Toutianoush A (2001) Eur Phys J E 5:29
81. Tieke B (2002) In: Tripathy SK, Kumar J, Nalwa HS (eds) Handbook of polyelectrolytes and their applications. American Scientific, chap 8
82. Schwarz HH, Richau K, Paul D (1991) Polym Bull 25:95
83. Schwarz HH, Scharnagl N, Behling RD, Aderhold M, Apostel R, Frigge G, Paul D, Peinemann KV, Richau G (1992) German Patent 4 229 530 0
84. Richau K, Schwarz HH, Apostel R, Paul D (1996) J Membr Sci 113:31
85. Shiratori SS, Rubner MF (2000) Macromolecules 33:4213
86. Steitz R, Jaeger W, v Klitzing R (2001) Langmuir 17:4471
87. McAloney RA, Sinyor M, Dudnik V, Goh MC (2001) Langmuir 17:6655
88. Fery A, Schöler B, Cassagneau T, Caruso F (2001) Langmuir 17:3749
89. Park SY, Barrett CJ, Rubner MF, Mayes AM (2001) Macromolecules 34:3384
90. Krasemann L (1999) Dissertation, Universität zu Köln
91. Urairi M, Tsuru T, Nakao S, Kimura S (1992) J Membr Sci 70:153
92. Tsuru T, Nakao S, Kimura S (1995) J Membr Sci 108:269
93. Jin W, Toutianoush A, Tieke B (2003) Langmuir 19:2550
94. Stair JL, Harris JJ, Bruening ML (2001) Chem Mater 13:2641
95. Balachandra AM, Dai J, Bruening ML (2002) Macromolecules 35:3171
96. Dai J, Balachandra AM, Lee JI, Bruening ML (2002) Macromolecules 35:3164
97. Lebedev K, Ramírez P, Mafé S, Pellicer J (2000) Langmuir 16:9941
98. Farhat TR, Schlenoff JB (2001) Langmuir 17:1184
99. Dubas ST, Schlenoff JB (2001) Macromolecules 34:3736
100. Donath E, Sukhorukov GB, Caruso F, Davis S, Möhwald H (1998) Angew Chem Int Ed Engl 37:2201
101. Qiu X, Leporatti S, Donath E, Möhwald H (2001) Langmuir 17:5375

Received October 2002

Adv Polym Sci (2004) 165:211–247
DOI 10.1007/b11271

Characterization of Synthetic Polyelectrolytes by Capillary Electrophoretic Methods

H. Engelhardt · M. Martin

Institute of Instrumental and Environmental Analysis, Saarland University,
PO Box 151 150, 66041 Saarbruecken, Germany
E-mail: iaua@rz.uni-saarland.de

Abstract Capillary electrophoresis has widely contributed to the analysis of biological poly-electrolytes like DNA or proteins. Their behavior in the different electrophoretic separation systems is well understood. However, this knowledge is rather limited in the case of synthetic polyelectrolytes. This paper discusses our results on how it is possible to gain information on size, shape, and mobility from capillary electrophoresis experiments. Capillary gel electrophoresis (CGE) is the most valuable method for size characterization of polyelectrolytes. Different separation principles and sieving media are presented and an optimization concept is evaluated. Capillary zone electrophoresis (CZE) and capillary isoelectric focussing (cIEF) can be used to study the influence of solvent properties such as ionic strength, buffer pH, and temperature on the size and surface charge of polycations, polybetaines, and polyampholytes in free solution. The pH dependence of the mobility can be used for the determination of charge neutrality of the surface. Systematic variations of structural parameters lead to significant changes in electrophoretic behavior. Even the formation of intermolecular associates is easily detectable by CZE. All results are in good agreement with orthogonal classical polyelectrolyte characterization methods which makes capillary electrophoresis a fast and efficient supplement to established methods.

Keywords Capillary electrophoresis · Capillary gel electrophoresis · Capillary isoelectric focussing · Polyelectrolytes · Polystyrenesulfonates · Polyvinylpyridines · Polycarboxybetaines · Polyampholytes

Abbreviations

$[\eta]$	intrinsic viscosity
λ	detection wavelength
μ, μ_0	electrophoretic mobility; mobility in free solution
μ^*	electrophoretic mobility in gel solution related to mobility in free solution
μ_r	electrophoretic mobility related to internal standard
BGE	background electrolyte
HDB	hexadimethrine bromide
M	molecular mass
PSS	poly(styrenesulfonate)
PVA	poly(vinyl alcohol)
PVBC	poly(vinylbenzyl chloride)
p(VPy)	poly(vinylpyridine)
R_g	radius of gyration
R_h	radius of hydration
T	temperature
U	applied voltage
c	concentration
c^*	threshold concentration
m	size selectivity (in Capillary Gel Electrophoresis)
pI	isoelectric point

These results have been reported, in part, at the following international symposia: HPCE 2000 (Saarbruecken, Germany), APCE 2000 (Hongkong, PR China), HPCE 2001 (Boston, MA, USA), ANAKON 2001 (Konstanz, Germany), HPCE 2002 (Stockholm, Sweden), ISC 2002 (Leipzig, Germany)

1
Introduction

Capillary electrophoresis (CE), the combination of the classical flat-bed electrophoresis with the instrumental potential of liquid chromatography, has found wide applications in the analysis of biopolymers like proteins and DNA. Separation of DNA by CE has been of paramount importance in the human genome project. It has been obvious, therefore, to study the application of CE to the analysis and characterization of synthetic polyelectrolytes. This report will summarize our recent research on the application of the various CE techniques with anionic, cationic, and ampholytic polyelectrolytes.

The concept of CE encompasses various separation systems. The fundamental separation principle is, of course, electromigration of the analytes in an electric field applied to both ends of the capillary. This active electrophoretic transport–responsible for the separation–is always superimposed on the passive electroosmotic flow (EOF), which contributes to the transport of the analytes, but not to their separation. This EOF strongly depends on the pH and the ionic strength of the buffer and on the surface properties of the capillary. In the easiest experimental set-up the capillary (50 μm to 100 μm inner diameter, 50–100 cm in length) is filled with the buffer solution. As the EOF is usually directed to the cathode the sample is introduced at the anodic end by suction, pressure, or by applying an electric field. In the general instrumental set-up the detector (standard UV-detection, tunable wavelength or diode-array detection) is located in the proximity of the cathodic buffer reservoir. The molecules are transported by the EOF to the cathode and migrate either with (coelectroosmotic: cations) or against (counterelectroosmotic: anions) the EOF. Depending on the vectorial contribution of the EOF slow anions (active migration slower than the EOF) are also transported to the detector. It is therefore possible to analyze in one run all cations (migrating in front of the EOF) and the slower anions. Faster anions can be detected by performing the separation with a reversed field or with capillaries in which the EOF is reversed by physical or chemical surface modification. A schematic representation of the system is depicted in Fig. 1. This standard technique is also called (free) capillary zone electrophoresis (CZE).

Other CE techniques in which the EOF contributes to the separation are micellar electrokinetic chromatography (MEKC), in which neutral analytes can be separated in the presence of a micelle-forming agent, almost exclusively sodium dodecylsulfate (SDS). Analyte transport occurs with the micelles; the separation principle is the partitioning of the analytes between the buffer solution and the micelles, hence a real chromatographic process. In capillary electrochromatography the capillaries are packed with chromatographic stationary phases and flow of the mobile phase is generated by applying an electric field. Surface properties of the stationary phase are responsible for the magnitude of the EOF. For further details monographs on CE [1–4] and CEC [5–7] should be consulted.

Polyelectrolytes and biopolymers have a constant charge-to-surface ratio and migrate therefore with almost identical velocity. Consequently they can-

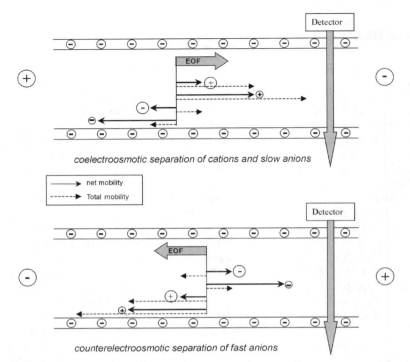

coelectroosmotic separation of cations and slow anions

Fig. 1 Modes of electrophoretic separation in fused-silica capillaries

not be separated by migration in the electric field. As the polymers differ in size, their electrophoretic migration can be hindered by filling the capillary with a gel, which is in most cases just a solution of a linear polymer. The separation range can be adjusted by the proper selection of the molecular weight of the "sieving" polymer and its molecular mass. Optimization strategies will be described in detail in this report. With this technique the EOF should be eliminated totally to avoid the gel from being swept from the capillary. Consequently the surface of the capillary has to be coated with a neutral, hydrophilic polymer, for example poly(vinyl alcohol) (PVA). The technique of capillary gel electrophoresis (CGE) with linear polyacrylamide (LPA) as sieving polymer contributed significantly to the analysis of the human genome [8].

Another electrophoretic technique–isoelectric focussing (IEF)–can also be transferred to capillaries [9]. In contrast to CZE, in cIEF amphoteric compounds are, because of their different isoelectric points, separated along a pH gradient formed in the capillary. Here also coated capillaries without EOF have to be used. The capillary is filled completely with a mixture of the amphoteric analytes and several ampholytes forming the pH gradient while applying the electric field. The ampholytes and the analytes take their position in the pH gradient formed. Gradient formation can be observed by monitoring the current in the capillary, which approaches a constant value when the gradient has been formed. After that the content of the capillary must be transported past the detector, either by applying pressure or by chemical mobilization. Capillary isoelectric focussing (cIEF) is a fast method of high efficiency for separation of ampholytic polymers.

2
Capillary Gel Electrophoresis (CGE)

2.1
Principal Considerations

The intrinsic problems of CE, e.g. influence of buffer concentration, problems with Joule heat dissipation, distortion of peak shape due to differences between salt concentrations in the separating buffer and analyte solution (electrodispersion), and effects on separation efficiency of analyte adsorption on the capillary surface have been widely discussed in handbooks on CE which may be used as references [1–4]. Low buffer concentrations are to be preferred to improve separation efficiency. A high salt concentration in the separation buffer counteracts analyte adsorption. Another important factor is elimination of the EOF and the stability of the surface coatings at high pH (pH>7.5). In the selection of sieving polymers the viscosity of their solutions should be low, to shorten the rinsing time while replenishing the buffer solution (the usual pressure for replenishing the capillaries with commercial instruments is mostly restricted to a maximum of 2 bar). Low buffer viscosity prevents bubble formation (heat dissipation) leading to current interruption. On the other hand, the concentration of polymer must exceed the so-called threshold concentration, c^*, above which entanglement of the independent molecular chains of the sieving polymer starts, which is necessary for efficient separation of polyelectrolytes. This is one important factor in selecting an appropriate water-soluble sieving polymer. Its molecular mass is important for adjustment of the sieving range.

The threshold concentration c^* can easily be determined with a CE instrument with good temperature and pressure control [10, 11]. In Fig. 2 the experimental curves for dextrans differing in molecular mass are given. The specific viscosity is determined via the time required to replace the polymer solution by the aqueous buffer. For time measurement a UV-absorbing neutral solute is injected. As the synthetic sieving polymers always have a molecular mass distribution entanglement of the chains of different lengths starts at different concentrations. Thus c^* is given more or less by a concentration interval rather than by a distinct point. Therefore a concentration at least twice as high as c^* (determined via the crossing point of the two tangents) seems appropriate in practice. The threshold concentration can also be calculated via an empirical equation using the intrinsic viscosity and other characteristic data of the polymer [12].

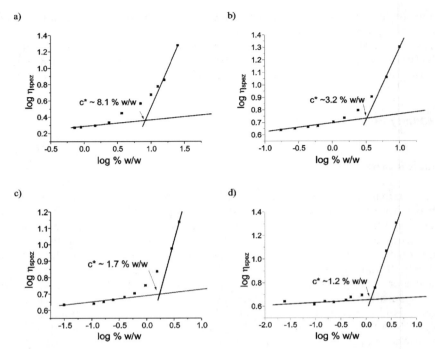

Fig. 2 Determination of the threshold concentration c^* for several dextrans as buffer additives: (**a**) dextran T10, (**b**) dextran T70, (**c**) dextran T500, (**d**) dextran T2000

For separations in CGE the "pore size" of such a polymer solution is of interest. The segment between two points of entanglement can be regarded as an independent sub-unit of the polymer, which can undergo the random walk per se. The volume enclosed by this chain segment is called a "blob" which can be used as the pore size of the sieving media [12]. In DNA analysis it has been shown that the "pore size" of an entangled polymer solution does not depend on the degree of polymerization but only on its concentration. The consequence is that two solutions of the same type of polymer of the same concentration but differing in molecular mass may have the same "pore size" as long as they are entangled. Because of differences in molecular mass the viscosities of the solutions are different. This is important in experimental application as low viscosity is desirable with CE instruments.

When plotting the logarithm of the relative mobility of a given polyelectrolyte versus the logarithm of its molecular mass (or chain length) a calibration curve as depicted schematically in Fig. 3 is obtained. In DNA analysis, the observed phenomena have led to the discussion of different migration models [13]. When the radius of gyration of the analyte is smaller than the "blob" size of the sieving polymer the migration behavior can be described by the Ogston model [14, 15]. Because of its low mass selectivity it is unimportant for molecular mass determination.

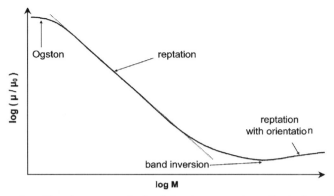

Fig. 3 Schematic calibration curve obtained for the separation of polyelectrolytes in entangled polymer solutions

The region with the highest mass selectivity is described by the so-called reptation model [16]. When the radius of gyration of the analyte is larger than the "blob" size of the sieving matrix the analyte molecule has to change its form from a coil to a stretched confirmation in order to be able to twist reptile-like through the entangled polymer network. For DNA molecules this movement has been visualized by fluorescence microscopy [17, 18].

The third region, where mobility reaches a minimum—mobility is independent of molecular mass—is also unimportant in practice, but is a focus of theoretical discussion [17, 19].

2.2
CGE of Synthetic Polyelectrolytes

CE techniques have great advantages over conventional chromatographic methods for characterization of polyelectrolytes, because in the pure aqueous system unwanted interactions with a stationary phase are excluded. However, strongly basic polyelectrolytes may adhere to the capillary surface, leading to peak distortion. In this case similar precautions have to be taken as in the separation of proteins and DNA, i.e. use of hydrophilically coated capillaries.

2.2.1
Analysis of Poly(Styrenesulfonates) (PSS)

As PSSs are fully dissociated even at low pH, they can be directly separated in this medium, because the EOF is strongly suppressed at pH values below 2.5. Wall adsorption of PSS has never been noticed [20]. Some authors discussed wall adsorption [21] because they had not been able to detect PSS at intermediate pH values (pH=5). However, with these conditions at pH 5 the migration of the PSS and the EOF may just be identical in opposite directions. When the separation has to be accomplished at higher pH values the EOF has to be suppressed by coating the capillary wall. A poly(vinyl alcohol) coating has been found to be optimal in respect of hydrophilicity, stability, and effectiveness of EOF suppression [20, 22] and is widely used in DNA analysis. No difference between the separation of PSS standards could be observed whether the separation was accomplished in uncoated capillaries at low pH (pH$\leq$2.5) or in coated capillaries at higher pH [19].

The optimization of PSS analysis in respect of type and concentration of the sieving polymers has been summarized in a recent publication in this series [19]. Most effective were solutions of dextrans, because they had excellent sieving properties at low viscosity at the required concentrations. Optimal discrimination of PSS according to molecular mass could be achieved in the molecular mass range up to 300,000 g mol^{-1} either with a 5% (w/w) solution of dextran T2000 or with a solution of 10% (w/w) dextran T500. With dextran T70 (15% w/w) mass discrimination was possible up to molecular mass of 150,000 g mol^{-1}. Poly(ethylene glycol) (20,000 g mol^{-1}) was not effective even at high concentrations.

Similar to the findings in DNA analysis the separation efficiency decreases with increasing field strength. Optimal conditions are around 200 V cm^{-1}. It has also been reported [11] that the mobility decreases with increasing ionic strength. Because the size of the PSS depends also on the buffer counterions, an influence of the counter-ions on mobility has been observed [11].

In a following section our results on optimization of the type of sieving polymer, its molecular mass, and its concentration will be described.

2.2.2
Analysis of Cationic Polyelectrolytes

Cationic polyelectrolytes are strongly adsorbed by the negatively charged surface silanols of the capillary. They are widely used (e.g. polybren) as buffer additives to reverse the EOF. Therefore, exclusively surface-coated capillaries have to be used in their separation.

In PVA-coated capillaries it was possible to separate at pH 2.5 the standards of poly-2-vinylpyridinium hydrochloride (p(2-VPy)) in the molecular mass range between 1500 and 1,730,000 g mol^{-1} with dextran T70 as sieving matrix [20]. An example is shown in Fig. 4, where a 5% solution of dextran T70 has been used. The efficiency of the monomolecular basic marker 4-aminopyridine is excellent, demonstrating the exclusion of secondary adsorptive effects at the capillary surface. Hence, the broad peaks of the polymeric standards are due to their polydispersity. As in CE the width of the peaks depends on their migration velocity through the detection window, no direct comparison of broadness of the individual peaks and analyte polydispersity is possible. However, for each individual peak the methods applied in SEC for calculation of the different molecular mass averages can be applied.

From the electropherograms of the p(2-VPy) standards calibration curves can be constructed as shown in Fig. 5. The curves are identical whether stan-

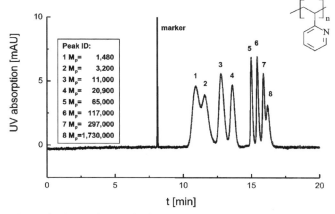

Fig. 4 Separation of p(2-VPy) standard mixture in 5% w/w dextran T70. Separation conditions: marker: 4-aminopyridine; capillary: PVA 50/57 cm, i.d. 75 μm; BGE: 0.05 mol L^{-1} phosphate pH 2.5+5% w/w dextran T70; U=+17.1 kV; injection for 10 s at 0.5 psi; T=25 °C; detection at 254 nm

dards are injected singly or as a mixture. This proves that analyte–analyte interactions are not present in this case.

When the averaged molecular masses determined by CGE are compared with those given by the suppliers (measured via SEC), molecular masses determined by SEC are always lower. When the CGE values are compared with results from MALDI–TOF measurements a closer correlation could be found. This may be an indication that in SEC polymer adsorption on the matrix was the reason for higher elution volumes, resulting in lower molecular masses. Of course, polydispersity is also affected by adsorption. Consequently CGE is a fast and reliable method for characterization of the molecular masses of polyelectrolytes and their distribution.

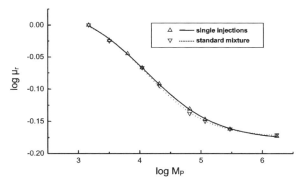

Fig. 5 Comparison of the calibration curves for p-(2VPy) in single injections and injection of the mixture. Separation conditions were as for Fig. 4

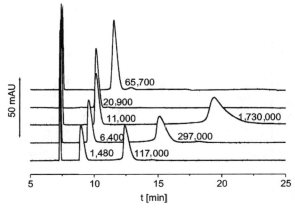

Fig. 6 Electropherograms of the standard test (p(2-VPy)). The numerical values represent the molecular mass M_P of the poly(2-vinylpyridine) standards (c=0.5 g L^{-1}). Separation conditions: marker: 4-aminopyridine; capillary: PVA 50/60 cm, i.d. 75 μm; BGE:
0.05 mol L^{-1} phosphate pH 2.5+1% w/w poly(vinylpyrrolidone) K90; U=+20 kV; injection for 15 s at 0.5 psi; T=25 °C; detection at 254 nm

The same standards can be separated with poly(ethylene glycol) (PEG
200; molecular mass 200,000 g mol^{-1}) in a 5% (w/w) solution or with
poly(vinylpyrrolidone) K90 (1% w/w). As can be seen in Fig. 6, the separation ranges and the broadness of the peaks are quite different, because of
differences in the sieving ranges of the polymeric additives.

2.2.3
Optimization of Sieving Polymer Selection

In DNA separations the molecular mass calibration curve of a CGE system is
usually described by double-logarithmic plotting of the reduced mobility μ^*
vs. molecular mass. The reduced mobility is calculated from the mobility μ
of the analyte in the sieving medium and that in free solution μ_0 ($\mu^*=\mu/\mu_0$).
The latter, however, is also influenced by other variables, e.g. the viscosity of
the buffer solution. On the other hand, it has been shown that the viscosity
has no direct relationship to the sieving potential of the polymeric additive.
In characterizing the potential of a sieving medium one is more interested
in mobility differences than in their absolute values. The addition of the
sieving polymer already influences the migration velocity of small (monomolecular) molecules. The migration time of a small molecule increases with
increasing polymer concentration as shown in Fig. 7 for the standard (4-
aminopyridine) with different polymeric additives.

Whereas mobility differences stay constant differences between migration
times increase with increasing separation time. This results in better resolution due to longer analysis times (corresponding to the use of a longer capillary), when increasing the concentration of the sieving polymer. Therefore,

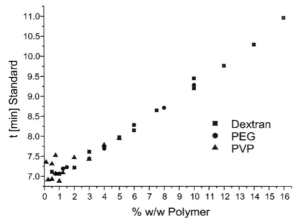

Fig. 7 Dependence on sieving polymer concentration of the migration time of the internal standard 4-aminopyridine in the p(2-VPy) standard mixture

the mobilities are better related to an internal standard, which should be a small molecule, not interacting with the sieving polymer, with similar ionic groups as the polyelectrolyte, and a higher mobility than the polyelectrolyte molecules. This relative mobility μ_r is calculated from the mobility of the polymer and that of the standard: $\mu_r=\mu_{(polymer)}/\mu_s$. The absolute values of μ_r are always smaller than unity and depend only on the mobility differences between the standard and the polymer in free solution. It is easy to recalculate these relative mobilities when systems with different standards are to be compared.

The calibration curve for p(2-VPy) can be constructed in the usual way by plotting the logarithm of the relative mobilities vs. the logarithm of molecular mass in the reptation range. The slopes of these calibration curves can be used as a measure for the size selectivity of the applied polymer additive. As sigmoidal curves are usually obtained, the slope demonstrates the efficiency of the separation system. Because the relative mobilities are ranged between zero and unity, negative values of the logarithm are always obtained, as demonstrated in Fig. 8. To characterize the efficiency of a separation system the slope at the inflection point of the calibration curve–the maximum sieving selectivity, m–can be used, as proposed by Dolnik [23, 24].

The range of the reptation region with the highest selectivity for a separation according to molecular mass is influenced by the type of the sieving medium, its molecular mass and distribution, and the concentration of the sieving polymer in the separating buffer. An additional but principal requirement for the selection of a sieving medium is its transparency in the UV region. Consequently dextrans and poly(ethylene glycols) are the first choice. As some polyelectrolytes can be detected at 254 nm

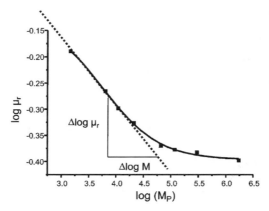

Fig. 8 Determination of the size selectivity, m, from the calibration curve of a polyelectrolyte with different peak masses M_P. m corresponds to the slope of the tangent at the inflection point

poly(vinylpyrrolidone) was also included in our studies. The polymers used are summarized in Table 1.

With all the sieving polymers used curves of very similar shape were obtained. As examples three different calibration curves are depicted in Figs. 9, 10, and 11. The curves themselves, and the inflection points, depend on the type of sieving polymer and its concentration. With increasing concentration the relative mobilities decrease. The highest decrease is observed in the region with the lowest molecular mass selectivity. This region extends to lower molecular masses, indicating a decrease in separation selectivity at higher concentrations of the buffer additives. The absolute value of the slope at the inflection point also increases with increasing concentration of the sieving polymer.

Table 1 Polymers investigated as sieving matrices

Polymer	Molecular mass (g mol^{-1}) [a]	Intrinsic viscosity, η (mL g^{-1})
Dextran T10	10,000	10.220
Dextran T70	70,000	25.885
Dextran T500	500,000	45.890
Dextran T2000	2,000,000	60.290
PEG 20	20,000	31.470
PEG 200	200,000	181.580
PEG 1000	1,000,000	558.000
PEG 8000	8,000,000	2,400.930
PVP K30	40,000	17.987
PVP K90	270,000	134.774

[a] data according to manufacturer

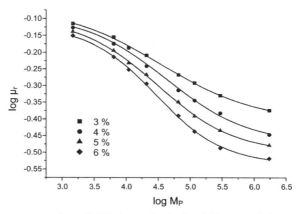

Fig. 9 Calibration curves for p(2-VPy) standards for different weight concentrations of dextran T500. Separation conditions: capillary: PVA 50/60 cm, i.d. 75 µm; BGE: 0.05 mol L^{-1} phosphate pH 2.5; injection for 15 s at 0.5 psi; T=25 °C; detection at 254 nm

In classic electrophoresis Ferguson demonstrated the linear dependence of analyte mobility on sieving polymer concentration in the region above the threshold concentration [25]. For constant concentration steps the decrease in mobility increases with increasing molecular mass. Similar to the so-called Ferguson plot the maximum sieving selectivity, m, can be plotted vs. the concentration of the sieving polymer in the buffer. An exemplary curve for poly(ethylene glycols) as sieving polymers for p(2-VPy) standards is depicted in Fig. 12. With increasing concentration of the buffer additive the sieving efficiency increases. The higher its molecular mass the more pro-

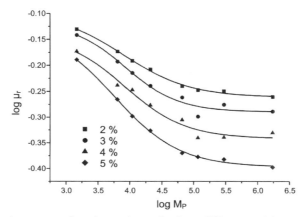

Fig. 10 Calibration curves for p(2-VPy) standards at different weight concentrations of PEG20. Separation conditions were as for Fig. 9

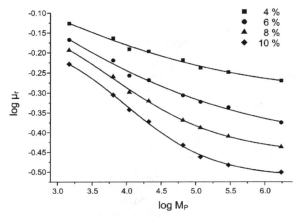

Fig. 11 Calibration curves for p(2-VPy) standards at different weight concentrations of poly(vinylpyrrolidone) K90. Separation conditions were as for Fig. 9

nounced this increase becomes. This means that the concentration of a high-molecular-mass sieving polymer can be much lower than that of a low molecular mass polymer to achieve identical selectivity. For the poly(ethylene glycol) system shown in Fig. 12 identical sieving selectivity (value of m) is achieved either with 0.1% PEG 1000 (MM 1,000,000 g mol^{-1}) or with a 3% solution of PEG 20 (MM 20,000 g mol^{-1}). The system resulting in the lowest viscosity should be optimal for CGE.

Different sieving polymers of identical molecular masses show, however, significant differences in their selectivity. This becomes obvious from Fig. 13, where the influence on selectivity for all concentrations and molecular masses for all the studied sieving media is summarized. It can be seen

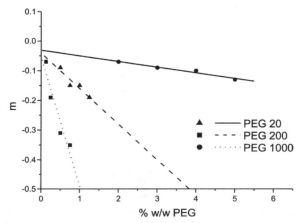

Fig. 12 Dependence of the size selectivity, m, on the concentration and molecular mass of the sieving polymer poly(ethylene glycol)

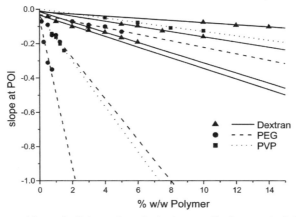

Fig. 13 Superimposition of all investigated sieving media in one Dolnik plot. Corresponding sieving polymers carry identical symbols and line styles

that the sieving selectivity of PEG and poly(vinylpyrrolidone) are superior to that of dextran of identical molecular mass. At a concentration of 0% no size selectivity is to be expected. As can be seen all curves approach a value of m close to zero. The small negative slope at a concentration of 0% can be explained by very small differences between the mobility of the standards in free solution (zone electrophoresis). Differences between the relative mobilities of the smallest standard (MM 1500 g mol^{-1}) and the largest (MM 1,730,000 g mol^{-1}) have been measured and are approximately 0.1.

2.2.4
Universal Calibration in CGE

The intrinsic viscosity $[\eta]$ is able to describe a relationship between the molecular mass and the size and shape of a polymer in solution. The logarithm of the radius of gyration shows a linear relationship to the logarithm of the product of molecular mass and intrinsic viscosity. This is applied in SEC to achieve the universal calibration according to Benoit [26]. This approach is also possible in CGE. By reducing the maximum sieving selectivity, m, for the intrinsic viscosity $[\eta]$ a linear relationship for all studied sieving polymers with their concentration in the buffer solution can be obtained, as shown in Fig. 14. It can be seen that all measured values are correlated in a linear graph. It can therefore be assumed that the sieving selectivity, m, of a polymer is directly proportional to its intrinsic viscosity. The sieving properties of a CGE system can thus be improved by increasing the molecular mass of the polymer. Improvements are also possible when a sieving polymer with a larger intrinsic viscosity is applied.

It has been reported in the literature that the sieving selectivity is highest when analyte and sieving molecules are of similar size (contour length) [27–29]. This could be proved by the following approach. The slopes of the cali-

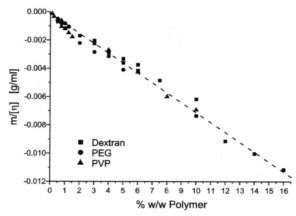

Fig. 14 Correlation between the concentration-dependent size selectivity, m, and the amount of buffer additive, c

bration curves vary with molecular mass. When plotting the first derivative of the sieving selectivity, m, vs. molecular mass, curves are obtained where the minimum shows the region of highest separation selectivity with this system. This is shown in Fig. 15 for dextrans as sieving media. With dextran T10 the optimum is below an analyte molecular mass of 1000 g mol^{-1} whereas for dextran T500 it is around 30,000 g mol^{-1}. The optimum sieving range increases with increasing molecular mass of the sieving polymer. With PEG 200 the optimal sieving selectivity is around 270,000 g mol^{-1} whereas for poly(vinylpyrrolidone) K90 an optimum around 150,000 g mol^{-1} is achieved. The valid general observation is that the sieving selectivity is highest where

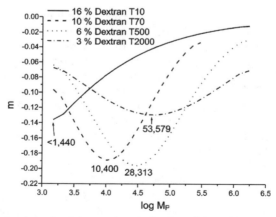

Fig. 15 Trend in the size selectivity, m, (1st derivative of the calibration curve) for different dextrans. The numerical values represent the molecular mass with the highest size selectivity

the radii of gyration of analyte and sieving molecules are identical. For dextrans—where the largest number of data is available—a linear relationship between the logarithm of intrinsic viscosity of the sieving polymer and the logarithm of the molecular mass of the analyte polymer could be obtained. The slope of this curve is 0.37. With the other sieving polymer studied values around 0.4 could be calculated for the relationship between both logarithmic values. From these observations it can be deduced that increasing the intrinsic viscosity of the sieving polymer leads to a shift of the maximum sieving selectivity to higher molecular masses. This is important in the selection of the proper sieving media. It may be added that in DNA separations the sieving selectivity of linear polyacrylamide used could be improved significantly by synthesizing polyacrylamide with extremely large molecular mass up to $9,000,000$ g mol^{-1} [30].

2.2.5
Extension of Sieving Range

In SEC the sieving range can be extended by coupling SEC columns packed with stationary phases differing in pore-size distribution. Optimal extended linear calibration curves are obtained when the average pore diameters of the coupled columns differ by a factor of 10 [31]. Mixing sieving polymers of different molecular masses should have the same effect in extending the range of separation efficiency in CGE. This is demonstrated in Fig. 16. A 3% solution of PEG 20 has the identical size selectivity, m, as a 0.1% solution of PEG1000, however, for different molecular masses. Mixing of the two solutions is not possible, because the individual concentrations would be reduced. However, by preparing the separation buffer with the two PEG siev-

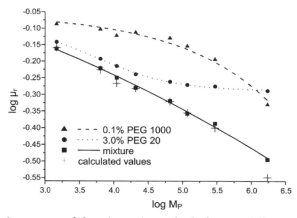

Fig. 16 Calibration curves of the p(2-VPy) standards for two different PEG types and the mixture of PEGs with different contour lengths. The *plus* symbol represents the theoretical curve for the mixture of sieving PEGs calculated from measurements on the single-PEG systems

ing polymers an extended calibration curve could be obtained. As can be seen in the figure the resulting calibration curve is an addition of the sieving properties of the two individual systems. If the calibration curves of the two systems are known, the resulting curve of the mixed system can be calculated from the relative mobilities of the analytes in both individual systems.

2.2.6
Conclusions

Optimization of polymer characterization and separation according to molecular mass can best be achieved by selection of the proper sieving polymer with an appropriate intrinsic viscosity. The average mass–or, better, contour length–has to be adjusted to that of the polymeric analyte. The maximum sieving selectivity increases with increasing intrinsic viscosity. The higher this increase with molecular mass, the stronger is its influence on the extension of the sieving selectivity to higher molecular masses. This means that the reptation range can be extended to higher masses by increasing the molecular mass and concentration of the sieving polymer. The concentration of the sieving polymer cannot, however, exceed a certain limit, because the viscosity of the buffer solution limits the replenishment time of the capillary and influences analysis time, which also should not exceed a certain limit. The UV transparency and the solubility in the buffer are also prerequisites in the selection of a sieving polymer.

The field strength should be as low as possible to improve the size selectivity, but not too low to limit analysis time. High field strength also gives rise to temperature problems. The influence of temperature on size selectivity is not so straightforward. With DNA and linear polyacrylamide increasing the temperature leads to an improvement in size selectivity [16]; with PSS and dextran the opposite has been reported [32].

2.3
Inverse CGE

So far UV-absorbing polyelectrolytes have been separated in UV-transparent sieving polymer solutions, i.e. the sieving polymer stays stagnant within the capillary and the analyte squeezes reptile-like through the "pseudo stationary phase" forced by the applied electric field. The fact that the maximum sieving selectivity is approached when both analyte and sieving polymer are closely related in size can be explained by interactions on a molecular basis, e.g. dispersion forces, hydrogen bonding, etc. The stronger these interactions over the whole contour length, the more the mobility of the complex will decrease. This is the principle of an inverse separation system, consisting of a migrating polyelectrolytic sieving matrix like poly(vinylsulfate) (sulfuric acid ester of poly(vinyl alcohol)) which carries the uncharged non-migrating polymer along the capillary. The separation depends on the degree of interaction between the analyte (uncharged) and matrix (charged). Of course, this is only possible at matrix polymer concentrations beyond the threshold concentration (for poly(vinylsulfate) an overlap threshold concentration of 0.17% *w/w* was determined after removal of surplus sulfuric acid by dialysis) [33].

Dextrans have been used as analyte molecules. For their detection they had to be derivatized with 7-aminonaphthalene-1,3-disulfonic acid (ANDSA). The contribution of the sulfonic acid groups to overall polymeric analyte migration can be neglected because the reagent is only reacting with the reducing sugar end-groups of the polymer. Only with this fluorescent label it was possible to detect even the large polymer molecules carrying only one label molecule. Due to the contribution of the introduced charges on the derivatized dextrans, especially the small oligomers, discussion of the observed separation is complex.

Dextran oligomers with a rather small molecular mass exhibit an intrinsic mobility due to the label. On the other hand, they are too small to interact with the sieving matrix and migrate therefore in a counterelectroosmotic way, as in free CZE; the smallest oligomer with the highest mobility is last to be transported by the EOF to the detector. This is shown as case A in Fig. 17.

Dextrans of medium molecular mass (e.g. around 21,000 g mol^{-1}) do not have any intrinsic mobility, the charge introduced by the label does not have any influence with an analyte of this molecular mass. Interaction with the sieving matrix does still not exist, consequently this molecule is transported by the EOF like an uncharged molecule in CZE (case B in Fig. 17).

Dextrans of high molecular mass also do not have an intrinsic mobility, but are large enough to interact with the migrating sieving medium. The longer the analyte chain, the more intensive the contact with the matrix, and the slower its transport to the detector in this counterelectroosmotic technique (case C in Fig. 17).

All these three cases can be found in the separation of dextrans and pullulans. The migration of the large polymer is, of course, overlaid by the small oligomers noticeable as "spikes" on the broad peak. The net mobilities of the derivatized pullulans and dextrans as a function of their molecular mass is shown in Fig. 18. In the oligomeric region the net mobility first decreases with increasing molecular mass, leading to a minimum at the velocity of the EOF (net mobility zero), and increases again when interaction with the migrating sieving medium (pseudo stationary phase) is possible and transport against the EOF takes place. The slope of the mobility curve increases with increasing matrix concentration due to the higher probability of analyte–matrix interactions. The pullulans as strictly linear polymers have a larger net mobility than the dextrans of identical molecular mass, due to the larger radius of gyration and better and stronger complex formation with the migrating sieving polyelectrolyte.

The interaction of the polymeric analyte molecules (charged or uncharged) with the polymeric sieving media (uncharged or charged) is a general and principal method for analytical separation of water-soluble polymers. Transfer of this technique to water-insoluble polymers by applying the technique of non-aqueous capillary electrophoresis (NACE) seems feasible.

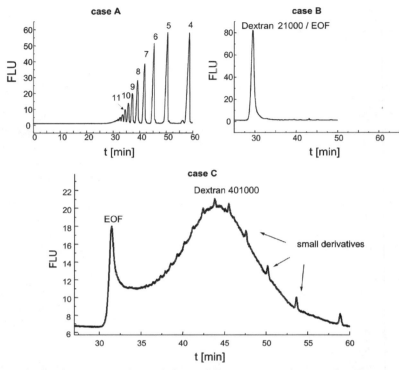

Fig. 17 Electropherograms obtained from different ANDS-dextran derivatives in the described separation system. Samples: **case A**: dextran 1,080 (c=1 g L^{-1}), the numbers represent the degree of oligomerization; **case B**: dextran 21,400 (c=10 g L^{-1}); **case C**: dextran 401,000 (c=20 g L^{-1}). Separation conditions: fused-silica capillary 88/100 cm, i.d. 50 µm; BGE: 0.05 mol L^{-1} phosphate pH 6.0+2% w/w poly(vinylsulfate); U=+25 kV; injection for 24 s at 200 mbar; detection: EX: 259 nm, EM: 442 nm

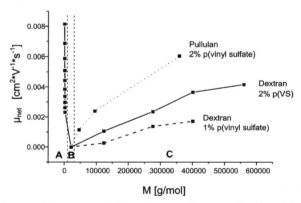

Fig. 18 Dependence of the net mobility of ANDS-derivatized polysaccharides on their molecular mass. Separation conditions were as for Fig. 17

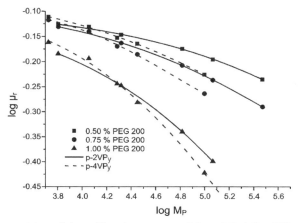

Fig. 19 Superimposition of the calibration curves of the p(VPy) for different concentrations of PEG200 as sieving medium

2.4
Application of CGE in Analysis of Polyelectrolytes

2.4.1
Comparison of the Mass Selectivity of Poly(2- and Poly(4-Vinylpyridines)

The optimization of the CGE systems has been accomplished with poly(2-vinylpyridine) (p(2-VPy)) standards [34]. The synthesis of standards of poly(4-vinylpyridines) (p(4-VPy)) is more problematic. Although the chemical structure is very similar, physical and chemical properties are different. It is of practical interest whether the calibration curves for p(2-VPy) can be used to characterize p(4-VPy). For the analysis identical separation systems could be used. Figure 19 compares the calibration curves for p(2-VPy) and p(4-VPy). The curves do not correlate. At low sieving polymer concentration the relatively small standards still migrate in the Ogston range. With increasing polymer concentration the curves move in the reptation range. The intercepts of both calibration curves also move to higher molecular masses. Nevertheless good mass selectivity could be achieved for p(4-VPy). The slope in the reptation region is even higher than with p(2-VPy). An explanation could be differences between the ternary structures of the molecules in solution.

2.4.2
Polyelectrolytes Based on Poly(Vinylbenzyl Chloride) (PVBC)

Polyelectrolytes are obtained from PVBC by reaction with ternary amines. Usually the polymers are characterized before quaternization, and the degree of substitution is determined via titration or NMR measurements. From

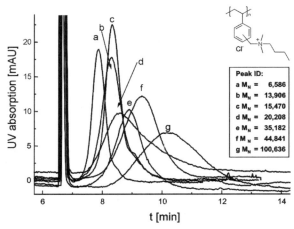

Fig. 20 Superimposed electropherograms of the singly-injected BM-PVBC in 0.75% *w/w* PEG200. Separation conditions: samples dissolved in 0.01 mol L^{-1} HCl; capillary: PVA 50/60 cm, i.d. 75 μm; BGE: 0.05 mol L^{-1} phosphate pH 2.5+0.75% *w/w* PEG200; *U*=+19 kV; injection for 15 s at 0.5 psi; *T*=25 °C; detection at 214 nm

the reaction products of PVBC with butyldimethylamine a variety of standards were available [34]. With these strongly basic analytes their adsorption at the capillary surface (despite coating with PVA) could not be totally excluded with the systems used for p-2VPy characterization. To minimize analyte adsorption flushing of the capillaries with a solution of 0.4% hexadimethrine bromide (HDB) in the buffer between the runs is essential. (Direct addition of HDB to the buffer is advisable when it is possible to work at higher detection wavelengths). An overlay of the electropherograms of the available standards with PEG as sieving matrix is shown in Fig. 20. From these runs calibration curves could be constructed [34]. With the low molecular mass samples only at higher PEG concentrations is the reptation regime approached. Comparing the calculated molecular mass averages and polydispersities from the precursor and those from the modified polyelectrolytes obtained by CGE a high correlation between SEC and CGE data could be determined for medium-sized samples [34].

2.4.3
CGE with Indirect UV Detection

Indirect detection techniques with UV-absorbing buffer components and non-absorbing analytes are widely applied in CZE of small molecules. To diminish electrodispersion, i.e. to achieve symmetrical peaks, the mobilities of analyte ions and background electrolyte should match closely [2]. In this case the sieving effect will dominate and peak broadening is only due to polydispersity. When mobility differences are large, electrodispersion rules

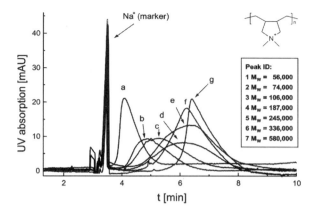

Fig. 21 Size-selective separation of p-DADMAC standards in 0.75% w/w PEG200. Separation conditions: capillary: PVA 20/30 cm, i.d. 75 μm; BGE: 0.009 mol L^{-1} pyridine pH 2.5 (adjusted with HCl)+0.75% w/w PEG200; U=+7 kV; injection for 20 s at 0.5 psi; T=25 °C; detection at 214 nm

over the sieving effect and leading ($\mu_{analyte} > \mu_{buffer}$) or tailing ($\mu_{analyte} < \mu_{buffer}$) peaks are obtained. In these cases another background electrolyte has to be chosen.

Poly(diallyldimethylammonium chloride) (pDADMAC) [35, 36] has low UV absorption. Using pyridine as background and PEG200 as sieving medium analysis of standards in the molecular mass range from 56,000 to 580,000 g mol^{-1} was possible, as shown in Fig. 21. As marker (standard for μ_r) the sodium ion was used. The mobilities of both background electrolyte and polyelectrolyte had been adjusted by CZE without any polymeric additive. As the sieving effect induces mobility differences, especially with the high-molecular-mass analyte, the asymmetric peak shape of this analyte (g) is certainly caused by an electrodispersion effect. Also the peak of the lowest molecular mass analyte (a) shows tailing. This is not caused by mismatch of mobilities (here a leading peak should appear) but can originate from increasing interaction with the sieving matrix with increasing molecular mass. From the relative mobilities (Na$^+$ as standard) the usual calibration curves can be constructed. As can be deduced from the figure, with the given sieving polymer (PEG200) and its concentration at molecular masses above 250,000 g mol^{-1} (e–g) no sieving effect is present. Other sieving polymers with higher molecular mass should be applied in this case. For the analytes with lower molecular mass (a–c) good correlation between SEC and CGE data could be observed [34].

2.4.4
Conclusion

CGE is suitable for the analysis of synthetic polyelectrolytes. Fast separations of high repeatability–when analyte interactions with the capillary wall are excluded–help to decrease separation time and solvent consumption compared with SEC. The only restriction is that CGE is only an analytical technique for small sample sizes. Selection of the optimal separation system with respect to molecular mass of the sieving polymer and its concentration is possible by applying simple rules and available data like intrinsic viscosity of potential sieving polymers. Non-UV-absorbing polyelectrolytes can be analyzed by applying indirect detection techniques. In various applications it has been shown that determination of molecular mass averages and molecular mass distribution is possible with CGE.

3
Capillary Free-Zone Electrophoresis of Synthetic Polyelectrolytes

The intrinsic electrophoretic mobility of polyelectrolytes is directly linked to the charge-to-surface ratio of the solvated molecule. Therefore CZE can quickly deliver valuable results on charge distribution. By using many chemically different types of polyelectrolytes, the influence of polyelectrolyte structure and of the type, pH, ionic strength and temperature of the background electrolyte can be investigated systematically.

One problem arises–interaction of positively charged polyelectrolytes with silanol groups on the capillary surface. Cationic polyelectrolytes are widely used in analytical CE separations for control and reversal of the electroosmotic flow in uncoated fused-silica capillaries. This becomes very disadvantageous in the measurement of polyelectrolyte mobilities themselves. As already discussed, unwanted electrostatic interactions between polyelectrolytes and the silanol groups of unmodified fused-silica capillaries can be suppressed by using capillaries coated with a hydrophilic polymer (e.g. poly(vinyl alcohol), PVA), but here hydrophobic interactions and hydrogen bonds can also lead to sample adsorption on the capillary wall. Very broad, tailing peaks will be the result in all cases, resulting in very poor determination of polyelectrolyte mobility. To tackle this problem, special coatings can be used. Alternatively–and sufficient in most cases–a selected choice of preconditioning rinsing steps between each separation, e.g. with highly concentrated polymer solutions (1% *w/w* linear poly(vinyl alcohol), 0.4% *w/w* hexadimethrine bromide) removing adsorbed sample residues and modifying the capillary surface by physisorption, will give much more stable results.

3.1
Influence of Structure on Mobility

The standard detection method (UV absorbance) requires UV-active chromophores in the analyte structures. A simple approach to the introduction of UV labels is realized with copolymers which combine UV-inactive and UV-absorbing structures. Hence, Clos [32] investigated poly(vinylamine)–polystyrene copolymers with significant UV absorption in the λ=254 nm region, due to the aromatic rings of the styrene blocks. These industrial block copolymers are synthesized by *grafting* of vinylformamide and styrene and are available in a broad variety of molecular masses and degrees of hydrolysis. Although these kinds of samples often tend to form micelles, because of alternation of the hydrophilic and hydrophobic parts in the polymer chain, the analytes performed well in CE systems. It could be demonstrated that samples with large and narrowly distributed degrees of hydrolysis gave very sharp peaks whereas the low-derivatized samples only result in flat, broad peaks. Hence, characterization and separation of different types of copolymer is possible.

Grosche and Martin investigated the influence of molecular structure on mobility for several polycationic and zwitterionic systems. In the case of poly(2-vinylpyridines) and poly(4-vinylpyridines) dependence of mobility on size was observed in the range 5300–257,000 g mol^{-1}; the dependence was smaller for the *ortho* compounds (Fig. 22) than for the *para* molecules (Fig. 23) [34]. However, these compounds can only be separated by CGE according to molecular size, as discussed above.

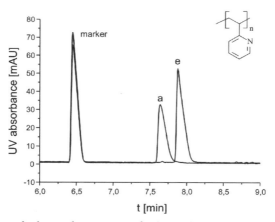

Fig. 22 Superimposed electropherograms of p(2-VPy) *a* (M_N=6,100 g mol^{-1}) and *e* (M_N=257,000 g mol^{-1}) in free solution. Separation conditions: sample (*c* =0.5 g L^{-1}) dissolved in 0.01 mol L^{-1} HCl, marker 4-aminopyridine; capillary: PVA 50/60 cm; BGE: 0.05 mol L^{-1} phosphate pH 2.5; *U*=+19 kV; injection for 5 s at 0.5 psi; *T*=25 °C; detection at 214 nm

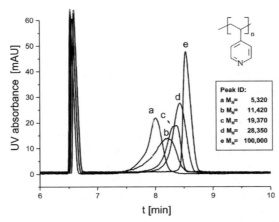

Fig. 23 Superimposed electropherograms of p(4-VPy) *a–e* in free solution. Separation conditions were as for Fig. 22

The analysis of permanently cationic polyelectrolytes based on quaternized poly(vinylbenzylchlorides) is hindered by the strong wall-interactions between the sample and the capillary surface. The positive charge of the analytes requires PVA-coated capillaries to prevent electrostatic interactions with the silanol groups, but even then adsorption occurs with increasing basicity of the molecules [37]. This problem could be solved only by the addition of 0.4% (*w/w*) HDB to the background electrolyte, as already discussed. By comparing the mobilities of different compounds, no explicit correlation between the structure and the migration behavior could be observed (Fig. 24). However, homologues of one chemical species show a systematic decrease in mobility with increasing length of side chains. This is not evident as the increase in hydrophobicity should result in a stronger collapse of the polymer coil and thus give smaller radii of gyration and higher mobilities. Most probably it is a diminished surface charge, due to a stronger shielding effect of the quaternized nitrogen atom, or formation of aggregates which is responsible for that effect. As with the poly(2-/4-vinylpyridines), size selectivity exists but is very small. In the molecular mass range from 6500 to 100,600 g mol^{-1} a non-linear decrease of mobility could be observed; it was insufficient to separate mixtures of these polyelectrolytes without size-discriminating mechanisms.

A more promising access to monitor structural influences is given with zwitterionic compounds like the polycarboxybetaines shown in Fig. 25. They are based on *N*-alkylcarboxylated poly(4-vinylpyridines) and poly(vinylbenzylchlorides) which carry one permanent cationic charge on the quaternized nitrogen atom and a pH-dependent anionic charge on the carboxyl group. All compounds form inner salts in bulk so that no effects due to counter-ion dissociation are to be expected. In all cases the separation system already developed for the cationic poly(vinylbenzylchlo-

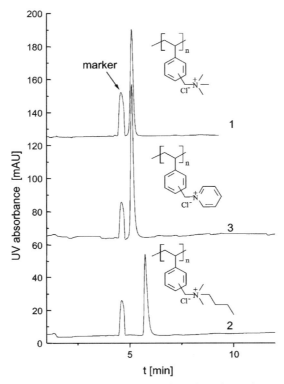

Fig. 24 Peak shapes and mobilities in phosphate/hexadimethrine bromide (HDB). Separation conditions: samples M-PVBC (**1**), P-PVBC (**3**), and BM-PVBC (**2**) dissolved in 0.01 mol L^{-1} HCl (c=0.5 g L^{-1}); BGE: 0.05 mol L^{-1} phosphate pH 2.5+0.4% w/w HDB; capillary: PVA 50/60 cm; U=+19 kV; injection for 5 s at 0.5 psi; T=25 °C; detection at 220 nm

rides) [34] was applied, including capillary dimensions, preconditioning steps, and field strengths. To speed up analysis the so-called "short-end injection" was used; in this the sample plug is introduced at the shorter outlet end of the capillary. Thus, the effective migration path length reduces from 20 cm to 10.2 cm which allows shorter analysis times and less contact with potential adsorption sites on the capillary surface. As background electrolyte, an HDB-free solution of 0.05 mol L^{-1} phosphate/0.05 mol L^{-1} acetate buffer at pH 2.0 was selected. The polybetainic substances generally show a much lower adsorption tendency at the capillary surface. A mixed buffer had to be used to achieve better reproducibilities than the pure phosphate buffer. The polycarboxybetaine mobilities at low pH showed significant differences in the mobility which cannot be explained only by differences in hydrophobicity. Injecting compounds of the same structure but varying molecular mass showed that the structure has a much larger impact on the migration than the molecular mass. It could be demonstrated that both the

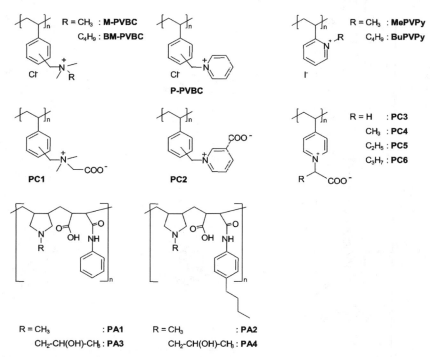

Fig. 25 Structures of selected polyelectrolytic compounds investigated by CZE

surrounding substituents and the flexibility and length of spacer groups between the cationic and anionic charges dominate the mobility. The main reason for changes in the mobility of polybetainic compounds is the degree of charge compensation which takes place between the cationic and anionic charges of the molecule. The more the carboxylic groups are able to form inner salts by association with the cationic nitrogen groups, the lower the overall mobility will be. This formation of associates is strongly affected by the shielding of the quaternary nitrogen atom or the flexibility of the spacer group to which the carboxylic group is attached. The stronger the spatial shielding effect of the nitrogen group or the more rigid the spacer chain, the less formation of associates takes place. Thus, the acidic groups can be protonated more easily and therefore are less able to compensate the positive charges, the driving force for electromigration in this system.

The nature of the association process (inter- or intramolecular) can be revealed by mixing compounds of the same structure but different molecular mass [34, 37]. Because the mobility of the investigated compounds is mass-dependent, with intramolecular association the electropherogram of the mixture should indicate superimposition of the two single injection peaks. If intermolecular associates are predominantly formed, one single, sharp peak should appear for the mixture. Both cases could be realized in practice:

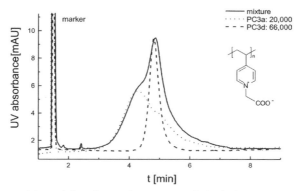

Fig. 26 Superimposition of the electropherograms of single injections and the mixture of compounds PC3a (M_W=20,000 g mol^{-1}) and PC3d (M_W=66,000 g mol^{-1}). Separation conditions: samples dissolved in 0.1 mol L^{-1} HCl; capillary: PVA 10/30 cm; BGE: 0.05 mol L^{-1} phosphate/0.05 mol L^{-1} acetic acid pH 2.0; U=+12 kV; injection for 5 s at 0.5 psi; T=25 °C; detection at 220 nm

Fig. 26 shows the electropherograms of PC3 with M_W=51,100 and 53,300 g mol^{-1}, and the mixture of both. The mixture correlates with the superimposition of the two different compounds so that intramolecular association can be assumed. In contrast, mixing two different types of PC2 leads to a sharp peak whose migration time lies between the migration times of the singly injected compounds; this is indicative of an intermolecular association mechanism (Fig. 27).

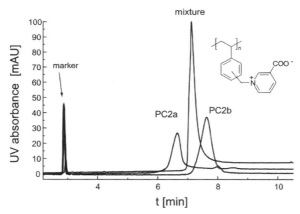

Fig. 27 Superimposition of the electropherograms of single injections and the mixture of compounds PC2a (M_W=53,000 g mol^{-1}) and PC2b (M_W=91,000 g mol^{-1}). Separation conditions: samples dissolved in 0.1 mol L^{-1} HCl; capillary: PVA 10/30 cm; BGE: 0.05 mol L^{-1} phosphate/0.05 mol L^{-1} acetic acid pH 2.0; U=+12 kV; injection for 5 s at 0.5 psi; T=25 °C; detection at 214 nm

3.2
Influence of pH on Mobility

In investigations of the pH-dependency of the mobility of samples PC1–PC6, the observations made so far could be confirmed. The series PC3–PC6 represents homologues with an increasing alkyl chain on the α-carbon atom which dominates the pK_a of the acidic group in accordance with the behavior of the related monomeric acids from acetic to valeric acid. As the mobility of the samples still varies at very low pH (pH<1.8), the background electrolyte had to be changed from phosphoric acid to a 0.05 mol L^{-1} hydrochloric acid/0.05 mol L^{-1} acetic acid solution. Furthermore, the field strength was reduced by a factor of 2 to diminish excessive Joule heating, due to better buffer conductivity. In addition, the ionic strength of the monovalent chloride buffer is less affected by pH variations than that of phosphate buffers.

A measured μ vs. pH diagram for the compounds PC3–PC6, PC1 and PC2 is depicted in Fig. 28. Three different regions can be distinguished. In the highly acidic pH range the mobility of all compounds can be considered as pH-independent, as expected for the case of full protonation of all carboxylic groups. As the influence of the acidity is eliminated, the observed mobility differences can only be attributed to different coil sizes of the polyelectrolytes. A systematic influence of the alkyl chain length appears with homologues PC3–PC6–the longer the chain, the higher the mobility. This can be explained by an increase in hydrophobicity which leads to a stronger collapse of the polymer and thus to larger charge-to-surface ratios. The influence of the chain length decreases with increasing numbers of methylene

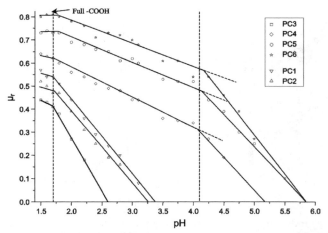

Fig. 28 μ vs. pH plot of compounds PC1–6. Separation conditions: samples dissolved in 0.1 mol L^{-1} HCl; capillary: PVA 10/30 cm; BGE: 0.05 mol L^{-1} hydrochloric acid/0.05 mol L^{-1} acetic acid; U=−2.5 k; injection for 5 s at 0.5 psi; T=25 °C; detection at 214 nm

groups. It is remarkable that the mobility remains constant at a pH<1.8. With increasing pH, dissociation of the acidic groups takes place and determines the hydrophobic effects. Weaker acids are deprotonated at higher pH values than stronger ones. This is expressed in the different slopes of the curves which also decrease with diminishing acidity. The linear trend permits the determination of that pH where the net charge of the polyelectrolyte approaches zero, by linear extrapolation of the curve to μ=0. For that purpose two major conditions have to be fulfilled–the pH dependence has to be constant in the whole pH range, and the number of anionic and cationic groups in the molecule has to be nearly identical. If there is a substantial excess of one charge type, the mobility approaches a constant level different from zero. Figure 28 shows the described results. The mobility of compounds PC1–PC3 decreases steadily with increasing pH. Extrapolation to μ=0 is possible, resulting in a zero net charge between pH 2.5 and 3.6. In contrast, the mobility graph shows a discontinuity around pH 4 for PC4–PC6. By extrapolating from the results in the range from pH 2–3.5, neutral points at pH>10 would be obtained, not explainable because the polycarboxybetaines do not have any basic functional groups. Responsible for this break is the pK_a value and the resulting buffer capacity of the analytes. With increasing pH the buffering effect of the weaker acids leads to a reduced decrease of mobility. When the buffer capacity is exhausted, mobility decreases rapidly. In principle, therefore, such measurements can be used for pK_a determination of unknown betaines.

3.3
Influence of Temperature on Mobility

In CE, an increase in temperature reduces buffer viscosity leading to higher mobilities and shorter analysis times. Unfortunately the buffer conductivity increases at the same time, resulting in higher currents and extended Joule heating. With increasing temperature the risk of poor reproducibility and/or the total collapse of migration due to bubble formation also increases. It is not, therefore, commonly used for CE optimization.

On the other hand, temperature also affects the coil size of polymers and polyelectrolytes. In general, the radii of gyration increase with increasing temperature, and the polymer coil will be enlarged. Assuming a constant charge of the polyelectrolyte, the consequence should be a reduction in mobility due to a smaller charge-to-surface ratio. This has been tested with two partially quaternized poly(2-vinylpyridines) MePVPy and BuPVPy (Fig. 25) in the temperature range from 15 to 50 °C. As only 70% of the pyridine groups are derivatized, a buffer of pH 7.0 was selected to exclude additional charges due to protonation of the aromatic nitrogen. With the standard separation system (PVA-coated capillary, phosphate buffer) reproducible separations could not be achieved. However, on using a mixture of 0.025 mol L^{-1} phosphate and 0.025 mol L^{-1} acetate buffers the expected increase of mobility could be observed for the 4-aminopyridine standard (Fig. 29). The mobility of the polyelectrolytes, however, increases also. With temperature-depen-

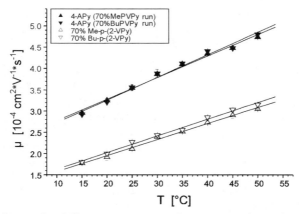

Fig. 29 Dependence of mobility on temperature for 70% Me-p(2-VPy) and 70%Bu-p(2-VPy). Separation conditions: samples: 1 g L^{-1} dissolved in water; capillary: PVA 10.2/31.2 cm, i.d. 75 μm; BGE: 0.025 mol L^{-1} phosphate, 0.025 mol L^{-1} acetate pH 7.0; $U=-5$ kV; Injection for 3 s at 0.3 psi; detection at 214 nm

dent light-scattering experiments it has been found that R_g and R_h increase, which should result in reduced mobility [38]. This might be because the degree of dissociation increases with increasing temperature. This effect overcompensates the volume extension of the polymer chain and thus leads to an effective amplification of the charge-to-surface ratio. In this context it should be mentioned that the type of buffer also leads to theoretical unpredictable mobility changes. In particular the mobility of MePVPy increases substantially on changing from chloride to acetate buffer; this can be related to greater collapse of the polymer in the more hydrophobic surroundings.

4
Capillary Isoelectric Focussing

Capillary Isoelectric Focussing (cIEF) is a method which separates samples according to their different isoelectric points (pI). This is an orthogonal principle to classical CZE which discriminates molecules according to their charge-to-surface ratio. cIEF has a very high separation efficiency, mainly because the usual band-broadening of the sample zones due to diffusion processes is strongly reduced by the separation mechanism itself. By the time a focussed sample molecule is leaving the sample zone due to diffusion, it drifts into the formed pH gradient and hence will carry again a net charge different from zero leading to a migration back into the focussed sample plug. Separation efficiency is very high while the limits of detection are improved because the whole capillary volume is filled with analytes prior to analysis which leads to high peak concentrations. In the case of polyelectrolyte analysis, two main restrictions have to be considered. Of course, only true ampholytic molecules can be investigated by cIEF. In addition, only compounds with significant UV absorption in the higher UV wavelength regions (λ>260 nm) can be investigated. To form a linear, stable pH gradient within the capillary a carrier ampholyte is added to the sample solution. These carrier ampholytes consist of a broad variety of isoelectric substances (commonly amino acids) which have strong UV absorption below 260 nm. In order to detect selectively sample molecules only, it is essential to measure UV absorption at 280 nm. For further details, monographs on cIEF should be consulted.

4.1
cIEF of Synthetic Polyampholytes

For investigation of the pH dependence of the mobility, four different samples PA1–PA4 could be used (Fig. 25). Focussing was performed in a PVA-coated fused-silica capillary. All samples were dissolved at 0.3 g L^{-1} in MilliQ water which contained 2% (v/v) Servalyte carrier ampholyte. After 12 min of focussing, the samples were transported towards the detector (λ=280 nm) by hydrodynamic mobilization. As shown in Fig. 30, all four samples could be separated in this system. By using two marker proteins, ovalbumin (pI=4.7) and β-lactoglobulin (pI=5.1), the isoelectric points of all compounds could be calculated in a pI range between 4.6 and 4.9. Additionally, the pH-dependence of the mobility was monitored in the same way as described for the polycarboxybetainic compounds (above). In the acidic range linearity was good for all compounds. By linear extrapolation to the point of net mobility zero, pI values were calculated. Both methods give a good correlation of the results (Table 2).

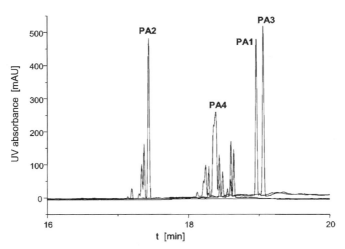

Fig. 30 Capillary isoelectric focussing of polyampholytes PA1–PA4. Separation conditions: capillary: PVA 40/50 cm, i.d. 75 µm; BGE: 300 ppm sample in MilliQ+2% Servalyt; anolyte: H$_3$PO$_4$, catholyte: NaOH; U=+30.0 kV; focussing time: 12 min; mobilization: 30 kV+0.1 psi; T=25 °C; detection at 280 nm

Table 2 Comparison of pI values obtained by CZE and cIEF for PA1–PA4

Compound	pI (CZE)	pI(cIEF)
PA1	4.8	4.6
PA2	4.9	4.9
PA3	5.0	4.6
PA4	4.7	4.7

In practice, both electrophoretic methods have their advantages. Capillary isoelectric focusing enables rapid and precise determination of isoelectric points but provides no information on the general pH-dependence of mobility or pK_a values. In addition, because of its high efficiency cIEF can reveal impurities which are not detectable under the broader peaks of CZE (Fig. 31). Capillary zone electrophoresis, however, determines pI values only indirectly; pK_a characteristics can also be determined but this takes longer.

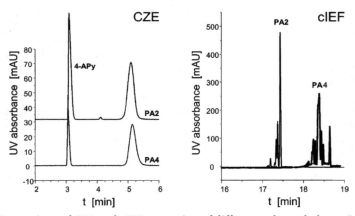

Fig. 31 Comparison of CZE and cIEF separation of different polyampholytes. Separation conditions were as for Fig. 28 and Fig. 30, respectively

5
Investigations of Counter-ion Dissociation of Sodium Poly(styrenesulfonates)

CE is a predestinate for study of polyelectrolyte counter-ion dissociation in free solution and study of the speed of this dissociation process. Counter-ion dissociation was studied with sodium poly(styrenesulfonates) of different molecular mass using indirect UV detection for quantification of sodium ions in the sample solution. The separation [39] was performed in an unmodified fused-silica capillary using a 0.005 mol L^{-1} imidazole buffer at pH 4.5 with potassium chloride as internal standard. In the electrophero-

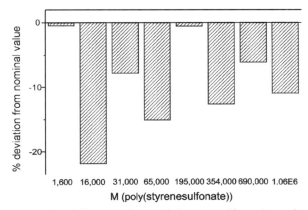

Fig. 32 Recovery rates for different sodium poly(styrenesulfonate) standards. Separation conditions: capillary: fused-silica 50/60 cm, 75 μm i.d.; BGE: 0.005 mol L^{-1} imidazole pH 4.5; U=+15 kV; Injection for 6 s at 0.5 psi; T=25 °C; detection at 210 nm

grams obtained the sodium cations give a sharp peak which indicates very rapid counter-ion dissociation and ion exchange with the buffer. Assuming slow exchange kinetics, a broad, tailing peak would be expected. With external calibration, the amount of sodium ions and the recovery rates for different PSS could be determined (Fig. 32). In all cases recovery rates were less than 100% of the theoretical values. However, an explicit trend with respect to the molecular masses of the samples obviously does not exist. Variations in the sodium content were most probably a result of contamination of the standards with sodium sulfate. To prove this a PSS sample (technical grade, MM 70,000 g mol^{-1}) was dialyzed with a cellulose membrane against pure water. The sodium content was determined both before and after dialysis using the CE method and compared with results obtained by atomic absorption spectroscopy (AAS) of the polyelectrolyte solutions (Table 3).

It can be seen that the CE and AAS results before dialysis differ strongly but reach a very similar level after purification by dialysis. One reason for these differences lies in the principle of CE itself. The demand for electroneutrality in the buffer system requires the replacement of the released sodium cations in the buffer and the polyelectrolyte environment by buffer cations, i.e. imidazolium ions. This replacement is an equilibrium reaction whose kinetics can lead to insufficient ion exchange during analysis, resulting in the observed variations. However, in all cases the recovery rate is no-

Table 3 Recovery rates of the sodium cations with different analytical methods before and after dialysis purification

Method	Prior to dialysis	After dialysis
CZE	66.6%	55.2%
AAS	85.0%	51.8%

ticeably distinct from 100%. Several reasons can be responsible. Either the PSS are not fully sulfonated (which is probable for technical grade material) or not all of the sulfonic acid groups contribute to the release of sodium ions. In addition, counter-ion replacement by protons formed by autoprotolysis or buffer electrolysis during the CE separation could also occur.

As discussed, quantification of counter-ion dissociation by means of CE is not trivial because reliable results are only obtainable under the following conditions:

- the rate of derivatization of the polyelectrolyte has to be known exactly (the best is 100%);
- the counter-ion content must be stoichiometric;
- during polyelectrolyte purification only excess salt ions should be removed; and
- cation replacement by protons should be excluded.

CE is, nevertheless, able to reveal valuable information about dissociation problems. The released counter-ions are replaced very rapidly by background electrolyte cations. Severe differences between the theoretically expected and experimentally determined counter-ion content are mostly caused by salt impurities from the synthetic process. In the example of PSS investigated the sodium content, which is substantially reduced by dialysis, can be attributed both to salt impurities and to ion exchange with protons of the buffer solution.

Acknowledgements Financial support by the DFG as part of the "Schwerpunkt Polyelektrolyte" is gratefully appreciated (EN 104/14–2). We especially thank the Fraunhofer Institut Angewandte Polymerforschung, Golm, Germany (W. Jaeger, J. Bohrisch, U. Wendler, T. Schimmel, K. Sander) and the Johannes-Gutenberg-University, Mainz, Germany (M. Schmidt, N. Volk, D. Vollmer) for cooperation and providing the polyelectrolyte samples. The results summarized here were extracted from the Ph.D. theses of Harald Clos (1998), Oliver Grosche (2001) and Markus Martin (2003).

References

1. Kuhn R, Hoffstetter-Kuhn S (1993) Capillary electrophoresis: principle and practice. Springer, Berlin, Heidelberg, New York
2. Engelhardt H, Beck W, Schmitt T (1994) Capillary electrophoresis: methods and potentials. Vieweg, Braunschweig
3. Weinberger R (1993) Practical capillary electrophoresis. Academic Press, San Diego
4. Kok WT (2000) Capillary electrophoresis: instrumentation and operation, Chromatographia supplement, vol 51. Vieweg, Braunschweig
5. Krull I, Stevenson RL, Mistry K, Swartz ME (2000) Capillary electrochromatography and pressurized flow capillary electrochromatography—an introduction. HNB, New York
6. Bartle KD, Myers P (2001) (eds) Capillary electrochromatography. Royal Society of Chemistry, Cambridge
7. Deyl Z, Svec F (2001) (eds) Journal of Chromatography Library: Capillary electrochromatography. Elsevier, Amsterdam
8. Hu S, Dovichi NJ (2002) Anal Chem 74:2833

9. Rodriguez-Diaz R, Wehr T, Zhu M (1997) Electrophoresis 18:2134
10. Bello MS, Rezzonico R, Righetti PG (1994) J Chromatogr A 659:199
11. Cottet H, Gareil P (1997) J Chromatogr A 772:369
12. Viovy JL, Duke T (1993) Electrophoresis 14:322
13. Slater GW (1997) In: Heller C (ed) Analysis of nucleic acids by capillary electrophoresis. Vieweg, Braunschweig
14. Ogston AG (1958) Trans Faraday Soc 54:1754
15. Tietz D (1988) In: Chrambach A, Dunn MJ, Radola BJ (eds) Evaluation of mobility data obtained from gel electrophoresis: strategies on the computation of particle and gel properties on the basis of the extended Ogston model, Advances in Electrophoresis, vol 2. VCH, Weinheim
16. Heller C (1997) (ed) Analysis of nucleic acids by capillary electrophoresis. Vieweg, Braunschweig
17. Schwartz D, Koval M (1989) Nature 338:520
18. Carlsson C, Larsson A, Jonsson M (1997) In: Heller C (ed) Analysis of nucleic acids by capillary electrophoresis. Vieweg, Braunschweig
19. Grosche O, Engelhardt H (2000) Adv Polym Sci 150:189
20. Clos HN, Engelhardt H (1998) J Chromatogr A 802:149
21. Poli JB, Schure MR (1992) Anal Chem 64:896
22. Engelhardt H, Cunat-Walter MA (1995) J Chromatogr A 716:27
23. Dolnik V (1998) Oral presentation at HPCE'98, Orlando
24. Dolnik V, Gurske WA (1999) Electrophoresis 20:3373
25. Ferguson KA (1964) Metabolism 13:985
26. Grubisic Z, Rempp P, Benoit H (1967) J Polym Sci Lett 5:753
27. Minarik M, Gas B, Kenndler E (1997) Electrophoresis 18:98
28. Barron AE, Soane DS, Blanch HW (1993) J Chromatogr A 652:31
29. Baba Y, Ishimura N, Samata K, Tsuhako MJ (1993) J Chromatogr A 653:329
30. Salas-Solano O, Carrilho E, Kotler L, Miller AW, Goetzinger W, Sosic Z, Karger BL (1998) Anal Chem 70:3996
31. Mori S, Barth HG (1999) Size exclusion chromatography. Springer, Berlin, Heidelberg, New York
32. Clos HN (1998) PhD thesis, Saarland University, Saarbruecken
33. Wendler U, Bohrisch J, Jaeger W, Rother G, Dautzenberg H (1998) Macromol Rapid Commun 19:185
34. Grosche O, Bohrisch J, Wendler U, Jaeger W, Engelhardt H (2000) J Chromatogr A 894:105
35. Dautzenberg H, Görnitz E, Jaeger W (1998) Macromol Chem Phys 199:1561
36. Wandrey C, Görnitz E (1995) Polym News 20:377
37. Bohrisch J, Grosche O, Wendler U, Jaeger W, Engelhardt H (2000) Macromol Chem Phys 201:447
38. Volk NH (2002) PhD thesis, University of Mainz
39. Beck W, Engelhardt H (1992) Chromatographia 33:313

Received October 2002

Author Index Volumes 101–165

Author Index Volumes 1-100 see Volume 100

Subject Index

T